GCSE in a week

Science

Kevin Byrne
Dan Evans, Jim Sharpe, Cathy Walters, Alex Watts &
Joanna Whitehead
Abbey Tutorial College
Series Editor: Kevin Byrne

Where to find the information you need

Cells and the movement of molecules	1
Plants and photosynthesis	7
Nutrition	13
Breathing and respiration	20
Blood and circulation	25
Nerves and hormonal coordination	31
Genetics	37
Ecology	43
The periodic table	49
Chemical equations	54
Atomic structure and bonding	61
Rates of reactions	67
Rocks and plate tectonics	73
Metals and the reactivity series	78
Non-metals	83
Chemicals from oil	89
Radioactivity	95
Energy/energy transfer	100
Energy resources	107
Light and the electromagnetic spectrum	112
Waves and sound	118
Electricity	124
Electromagnetism	129
Forces and motion	135
The Earth and beyond	141
Mock exam	146
Answers	154

Letts Educational
Aldine Place
London W12 8AW
Tel: 0181 740 2266
Fax: 0181 743 8451
e-mail: mail@lettsed.co.uk
website: http://www.lettsed.co.uk

Every effort has been made to trace copyright holders and obtain their permission for the use of copyright material. The authors and publishers will gladly receive information enabling them to rectify any error or omission in subsequent editions.

First published 1998
Reprinted 1998, 1999

Text © Kevin Byrne, Dan Evans, Jim Sharpe, Cathy Walters, Alex Watts
and Joanna Whitehead 1998
Design and illustration ©: BPP (Letts Educational) Ltd 1998

All our Rights Reserved. No part of this publication may be reproduced, stored in a retrieval system, or transmitted, in any form or by any means, electronic, mechanical, photocopying, recording or otherwise, without the prior permission of Letts Educational.

British Library Cataloguing in Publication Data
A CIP record for this book is available from the British Library.

ISBN 1 85758 702 2

Printed in Great Britain by Sterling Press, Wellingborough NN8 6UF

Letts Educational is the trading name of BPP (Letts Educational) Ltd

Cells and the movement of molecules

Test your knowledge

1 All living things move, respire, show sensitivity, grow, _reproduce_, excrete and need nutrition.

2 The basic unit of any living organism is the _cell_. Groups of cells of the same type are called _tissue_.

3 There are two main types of cell: animal cells and _plant_ cells.

4 A plant cell contains chloroplasts which contain a green pigment called _chlorophyll_.

5 The kidneys remove _waste products_ and control the amount of _water_ in the body. The lungs take in _oxygen_ and give out carbon dioxide.

6 Plant roots absorb water and _minerals_ into the plant.

7 A sperm cell is specialised because it has a _tail_ to swim towards the egg cell for fertilisation.

8 _Diffusion_ is the movement of molecules from a high concentration to a low concentration.

Answers

1 reproduce 2 cell / tissues 3 plant 4 chlorophyll
5 poisonous/toxic waste substances / water / oxygen/air
6 minerals 7 (long) tail 8 Diffusion

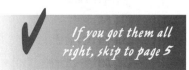

If you got them all right, skip to page 5

Cells and the movement of molecules

Improve your knowledge

1 All living things (animals/plants) do these seven things. This can be remembered as **MRS GREN**. Each letter represents a process.

Movement, **R**espiration, **S**how **S**ensitivity, **G**rowth, **R**eproduction, **E**xcretion and **N**utrition.

2 Groups of cells of the same type are called **tissues**. Tissues that work together to perform a function (job) are called **organs**. Organs work together as part of an organ system, e.g. the alimentary canal: stomach, small intestine, large intestine, liver and pancreas. The alimentary canal digests large insoluble particles to small soluble particles. An **organism** can either be made of one cell or even millions of cells, e.g. a bacterium or an oak tree.

A **light microscope** is used to study cells because it magnifies them (i.e. makes them bigger). Learn how to label the microscope: eyepiece lens, focusing knob, handle, (low and high power) objective lens, stage, mirror and stage clips.

3 There are two main types of cell: an animal and a plant cell. They contain different structures found inside the cell.

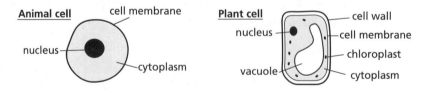

4 Learn the structure and function of each part of a plant cell:
Nucleus controls the cell.
Cell membrane allows certain substances into and out of the cell.
Cytoplasm is where chemical reactions take place and is the liquid part of the cell.
Cell wall holds up (supports) the plant and is made of cellulose.
Vacuole stores the sugar solution.

Chloroplasts contain the green pigment chlorophyll, which captures the sunlight for photosynthesis. Photosynthesis makes starch.

5 Each organ in animals has a particular function:

Brain controls the body.

Stomach digests food.

Small intestine digests food into smaller particles and absorbs the food particles into the bloodstream.

Large intestine absorbs water from food and stores the faeces.

Liver breaks down (detoxifies) alcohol, produces heat and stores glucose as glycogen.

Pancreas produces the hormone insulin, which stores glucose as glycogen in the liver.

Lungs take in oxygen (air) and get rid of carbon dioxide.

Heart pumps blood around the body.

Kidneys remove poisonous waste substances and control the amount of water in the body.

Bladder stores urine and gets rid of urine.

Testes produce the male sex cells (sperm) for reproduction.

Ovaries produce the female sex cells (egg cells or ova) for reproduction.

Eyes detect light.

Ears detect sound.

6 Each organ in plants has a particular function:

Flowers attract insects and produce seeds for reproduction.

Leaves photosynthesise to make food (starch) using sunlight.

Roots absorb water and minerals from the soil.

Stem transports water to leaves and transports sucrose to the growing areas of the plant.

7 There are some cells in animals and plants which are specialised to perform a particular job (function).

In animals

Egg cell (ovum)
It has a thin membrane to make it easy for sperm to get through and a layer of jelly to prevent damage.

Sperm cell
It has a long tail to swim to the ovum for fertilisation. The head contains genetic material that will decide your characteristics. It also contains enzymes to break down the layer of jelly around the egg cell.

Nerve cell (neurone)
It has a long axon to carry messages (as electrical impulses) around the body.

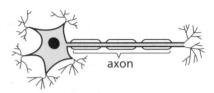
axon

Red blood cell
It does not contain a nucleus so that it has a larger surface area to carry oxygen around the body.

surface view   side view

In plants

Palisade mesophyll cell
It has many chloroplasts containing chlorophyll, which is needed for photosynthesis.

Root hair cell
It has a larger surface area to absorb water and dissolved minerals.

8. Molecules always spread themselves out evenly to fill all the available space. Molecules move from a region where there are a lot of them (i.e. concentrated) to regions where there are few of them (i.e. less concentrated) until the concentration becomes the same. This is called diffusion. The cell membrane is semi-permeable, i.e. small molecules like water, carbon dioxide and oxygen can pass through easily and larger molecules like proteins cannot pass through easily. Diffusion depends on temperature, pressure, concentration gradient, surface area and thickness of the membrane.

✓ *Now learn how to use this knowledge*

Cells and the movement of molecules

Use your knowledge

1 State three differences between cell **A** – plant cell (palisade mesophyll cell) and cell **B** – animal cell.   *Hint 1*

cell A    cell B

2 What is the function of cell **A** and how is it well suited for its function?   *Hint 2*

3 In which layer of the leaf is cell **A** found?   *Hint 3*

4 The stomach is part of the digestive system.   *Hint 4*

Name three other parts of the digestive system. _____ , _____ and _____ .

What does the digestive system do?
_____
_____

Hints and answers follow

Cells and the movement of molecules

1. What structures can you see in cell **A** compared to cell **B**?

2. Why are there many chloroplasts containing chlorophyll?

3. Is it the upper or lower surface of the leaf?

4. Which organs help to digest food?
   Look at numbers 2 and 5 in *Improve your knowledge*.

Answers

1 cell A (plant cell) has chloroplasts, a vacuole and a cellulose cell wall which are not found in cell B (animal cell) 2 to make starch / it has many chloroplasts for photosynthesis 3 palisade mesophyll layer found in the upper surface of the leaf 4 small intestine / large intestine / pancreas / liver / oesophagus or mouth / digests large insoluble food particles to small food particles

Plants and photosynthesis

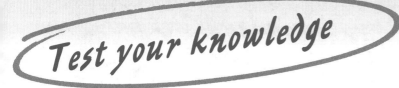

1. Photosynthesis is the process where a plant uses _____ energy to change carbon dioxide and water into _____ and oxygen.

2. The stomata are the _____ which allow gases in and out of the leaf.

3. Factors like light intensity, carbon dioxide and temperature are said to be _____ because they stop the rate of photosynthesis increasing.

4. Plants photosynthesise during the _____ .

5. Glucose is changed into starch, which is stored in _____ _____ .

6. The two transport tissues are called _____ and phloem. One carries water and dissolved minerals and phloem carries _____ .

7. _____ is the loss of water from the surface of the leaf by evaporation.

8. A plant which lacks nitrogen has _____ growth and _____ leaves.

9. A plant grows in response to a stimulus and this is called a _____ .

10. Phototropism is caused by a chemical called _____ .

Answers

1 solar / glucose/starch 2 pores 3 limiting 4 day
5 storage organs 6 xylem / sugar 7 transpiration
8 stunted / yellow 9 tropism 10 auxin

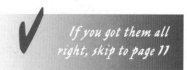

If you got them all right, skip to page 11

Plants and photosynthesis

Improve your knowledge

1 **Photosynthesis** is the process where a plant changes sunlight energy into chemical energy.

$$\text{carbon dioxide} + \text{water} \xrightarrow{\text{light energy trapped by chlorophyll}} \text{glucose} + \text{oxygen}$$

2 The structure of the leaf is adapted for photosynthesis. It is thin and flat so it has a large surface area to absorb sunlight and allow for gaseous exchange. Learn to label the cross-section of a leaf:

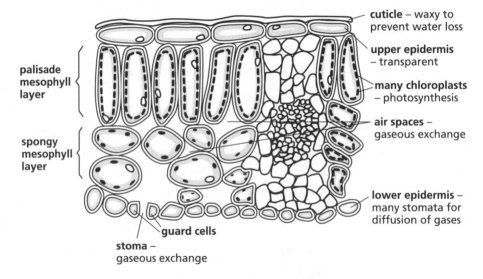

The **cuticle** is waxy to prevent water loss and protect the leaf. Plants which live in dry conditions have a thick cuticle.

The **palisade mesophyll cells** have many chloroplasts which contain chlorophyll to trap sunlight energy and make glucose.

The **spongy mesophyll cells** have many air spaces to allow gaseous exchange.

The **lower epidermis** has many stomata to allow gases to diffuse in and out of the leaf.

The **stomata** are the pores, which allow gaseous exchange. The size of stomata can be changed by guard cells.

3. The rate of photosynthesis can be increased by increasing the carbon dioxide concentration, light intensity and temperature. However, it increases only up to a maximum level because another factor slows the process. This factor is called a **limiting factor**.

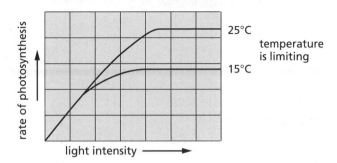

4. Just like animals, plants respire to release energy.

5. Glucose can be:
   a) used in respiration to release energy
   b) built up into cellulose for new cell walls
   c) built into proteins for healthy growth
   d) transported to storage organs and converted to starch.

   The starch is:
   a) stored in storage organs or leaves
   b) insoluble (it does not dissolve in water)
   c) changed back into glucose.

6. Water and dissolved minerals enter the plant by the root hairs and travel up microscopic tubes called **xylem** to the leaves. Water is used for photosynthesis and also gives the plant strength to stand upright. The **phloem** carries sugar (made by photosynthesis) to the growing parts of the plant or to be stored as starch.

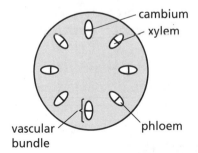

Cross-section through a stem

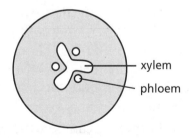

Cross-section through a root

Now learn how to use this knowledge

7 **Transpiration** is the loss of water from the leaves by evaporation and through the stomata by diffusion. The rate of water loss can be measured using a **potometer**.

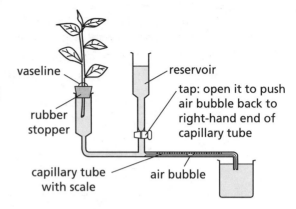

Factors which affect the rate of transpiration are:
a) **Wind** – more wind, faster the rate of transpiration (increased evaporation).
b) **Humidity** – more humidity, slower the rate of transpiration (decreased evaporation).
c) **Temperature** – higher temperature, greater rate of transpiration (increased evaporation).

8 A plant needs **mineral ions** to stay healthy. They are dissolved in the water and taken up by the root hair cells into the xylem vessels.

Mineral ion	Function	Effect if it is absent
Nitrogen	Forms part of proteins	Stunted growth; yellowing of leaves
Magnesium	Forms chlorophyll	Yellowing of leaves
Potassium	Making proteins; part of enzyme	Yellowing on edges of leaves

9 Plants can detect changes in their environment (stimuli) but they respond very slowly. Plant responses are related to the direction of the stimulus. This growth is called a **tropism**. The shoot grows towards light and this is called **positive phototropism**. If a seed is planted, the shoot grows upwards and the root grows downwards. The response to gravity is called **geotropism**. The shoot's growth is **negative geotropism** and the root's growth is **positive geotropism**.

10 Phototropism is caused by a chemical called **auxin**. This is made in the tip of the stem and is a **growth hormone**. It makes the plant cells grow faster.

Plant hormones can be used to:
a) encourage growth of roots in cuttings.
b) develop fruits rapidly for farmers.
c) kill weeds by making them grow out of control.

Plants and photosynthesis

Use your knowledge

20 minutes

An experiment was carried out to show gaseous exchange in leaves. The test tubes were set up as below and were left in the sunlight for 5 days.

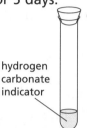

hydrogen carbonate indicator

covered in aluminium foil

Hydrogen carbonate indicator was used because it changes colour with different pHs.
- It is orange at neutral (pH 7).
- It is yellow in acid (pH below 7).
- It is red/purple in alkali (pH above 7).
- Carbon dioxide is an acidic gas and oxygen is neutral.

The results are shown here:

Tube	Colour of indicator
1	orange
2	purple/red
3	yellow

1 What changes in pH have happened in tubes 1 to 3? *(Hint 1)*

2 Explain what has happened in tube 2 to make the indicator turn purple/red. *(Hints 2/3)*

3 Explain what has happened in tube 3 to make the indicator turn yellow. *(Hints 4/5/6)*

4 What was tube 1 set up for and why? *(Hint 7)*

✓ *Hints and answers follow*

Plants and photosynthesis

Hints

1. What colour has the indicator turned? Use the information given to decide if the pH is neutral, acidic or alkaline.

2. Which gas has been used up from the air in daylight?

3. The leaf is photosynthesising. Which gas is used for this process?

4. Tube 3 is in darkness, therefore no photosynthesis can occur. What process occurs during the day and night?

5. Look at number 4 in *Improve your knowledge.*

6. What acidic gas is produced by respiration?

7. What is the test tube called that is set up to compare results?

Answers

1 tube 1 – no change / tube 2 – increase in pH / tube 3 – decrease in pH **2** The leaf has been in the sunlight and photosynthesising; it has used up the carbon dioxide in the air for photosynthesis, which has led to an increase in pH **3** The leaf has been in the dark. Therefore no photosynthesis has occurred. It has only been respiring and has produced carbon dioxide. This has made the solution turn acid and change to a yellow colour **4** Tube 1 was set up as a control to compare results and see if any other factor had changed the colour of the indicator solution

Nutrition

Test your knowledge

1 A balanced diet for humans includes proteins, _____ and lipids.

2 _____ solution is used to test if a food contains simple sugars. When this solution is added to food, it turns _____ _____ in colour if simple sugars are present.

3 Digestion is the breakdown of large, _____ molecules into _____ soluble molecules. For example, lipids in food are broken down into _____ _____ and _____ .

4 After being chewed in the mouth, food passes down the _____ to the stomach. The partly digested food then enters the _____ _____ . Food moves through the digestive system by contraction of muscles within the gut wall. This process is known as _____ .

5 _____ digestion occurs mainly in the mouth, where the teeth break the food into small pieces. This makes _____ digestion using enzymes faster. Enzymes _____ _____ chemical reactions in the body. For example, _____ breaks down carbohydrates to maltose.

6 The wall of the small intestine is folded to form thousands of _____ which increase the _____ _____ to absorb food. Each one of these has a wall only _____ cell thick and contains numerous _____ to carry food to the liver.

7 Some of the food absorbed after digestion is used in _____ to release energy. The rest of this food is used in _____ to build up the structures of the body.

Answers

1 carbohydrates **2** Benedict's / brick red **3** insoluble / smaller / fatty acids / glycerol **4** oesophagus / small intestine / peristalsis **5** Mechanical / chemical / speed up / amylase **6** villi / surface area / one / capillaries/blood vessels **7** respiration / assimilation

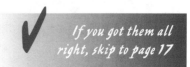

If you got them all right, skip to page 17

Nutrition
Improve your knowledge

1 The table below shows the different food groups (or nutrients) needed. A **balanced diet** has all these food groups in the right amounts.

Food group	Good sources	Uses	Effect if not eaten in diet (deficiency disease)
Proteins	Eggs, meat	Muscles, enzymes Energy supply	*Kwashiorkor* – bloated belly, slow brain development
Lipids (fats and oils)	Butter, milk	Energy store Heat insulation	
Carbohydrates (sugars)	Potatoes, pasta	Energy supply Plant structure	
Vitamin C	Citrus fruit (e.g. oranges), milk	Repair of damaged tissues Healthy teeth and gums	*Scurvy* – poor healing of wounds, bleeding of gums
Vitamin D	Butter, egg yolk	Calcium and phosphate ions uptake from gut Calcium depositing in bones	*Rickets* – weak and/or deformed bones
Calcium	Milk and cheese	Major part of teeth and bones	Fragile teeth and bones
Iron	Liver, egg yolk	Haemoglobin in red blood cells	*Anaemia* – lack of red blood cells
Fibre	Fresh vegetables	Movement of food through the gut	
Water	Drinking, fruit and vegetables	Animal bodies are over 75% water	

2 Chemical tests are used to see which type of carbohydrates different foods contain.

What are you testing for?	Chemical test used on food	Result if carbohydrate is present
Simple (reducing) sugars	Add blue Benedict's solution to food and warm	Solution turns brick red in colour
Starch (complex sugar)	Add brown iodine solution to food (must be cold)	Solution turns blue/black in colour

3 Digestion is the breakdown of the large, insoluble molecules in food into the smaller soluble molecules that make them up (basic units), so they can then be absorbed across the cells of the gut into the body and used for respiration or growth.

4 There are two types of digestion:
  a) **Mechanical digestion** – chewing in the mouth using teeth and churning by the muscles of the stomach breaks food into small pieces increasing its surface area, making further digestion easier.
  b) **Chemical digestion** – enzymes secreted by cells in the digestive system hydrolyse (break down) large food molecules into small soluble ones. These can be absorbed and transported to where they are needed. For example, starch, a carbohydrate, is broken down to maltose by the enzyme **amylase**.

5 The digestive system (alimentary canal/gut) breaks down the food ingested (taken in) and egests (removes) any indigestible material (e.g. fibre). The diagram below shows the layout of the human gut.

The human digestive system

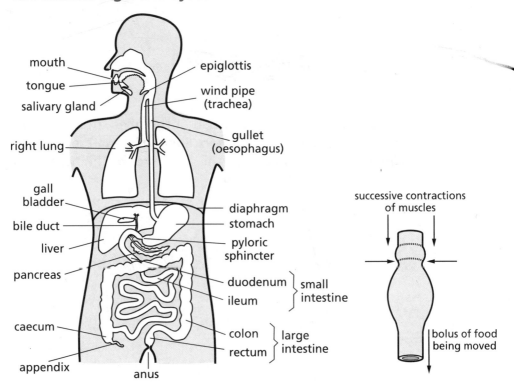

Peristalsis is the way food is moved along the gut. Rings of muscles in the wall of the gut contract behind the food, squeezing it along.

6. The small intestine is well suited to carry out its job of absorbing fully digested food into the blood:

   a) Thousands of finger-like projections called **villi** (singular **villus**) stick into the cavity of the intestine. Each villus is covered in tiny projections called microvilli. This means that there is a very large surface area for absorption.
   b) Each villus contains numerous blood capillaries for transporting absorbed food.
   c) The wall of each villus is only one cell thick, so there is only a short distance for the food to travel into the blood.

A longitudinal section of a villus in the small intestine

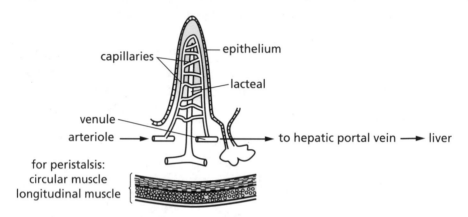

7. The small, soluble molecules absorbed at the small intestine are carried to the liver in the hepatic portal vein. Some of these are built back up into larger molecules to make up the structures of the body. This process is known as **assimilation**.

Now learn how to use this knowledge

Nutrition

Use your knowledge

The table below shows five different snacks and statements relevant to different food groups.

	Meal	Statement
A	Scrambled eggs on toast	Stops the disease scurvy
B	Glass of milk	Produces healthy teeth and bones
C	Spaghetti	Body building and provides iron
D	Glass of orange juice	Good energy food
E	Buttered bread	Heat insulation and energy supply

1 Pair up each of the meals above with the most appropriate statement.

2 A group of students were given the five meals above and asked to find out whether they contained starch or simple sugars. Explain how the students could do this.

3 People who do a lot of physical exercise often believe that they need a diet containing large amounts of protein. Suggest why this belief may be wrong.

4 In the last century, sailors on long sea journeys often suffered from bleeding gums and found injuries, such as cuts, didn't heal properly. What was the cause of the sailors' illness? Suggest what they could have done to improve their health.

The diagram shows the human digestive system.

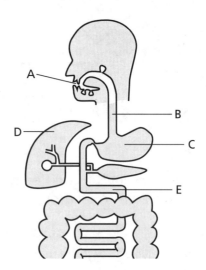

5 Name the parts labelled A to E.

6 Label on the diagram with the letter F, one position where the enzyme amylase is produced. Would amylase be involved in the digestion of bread or meat? *(Hint 5)*

7 What is the main role of the region E? State two characteristics of E which allow it to efficiently carry out this role. *(Hint 6)*

Some animals can develop a disease of the gut which causes the villi in the small intestine to be destroyed.

8 How will this affect the amount of food assimilated in the body? *(Hint 7)*

9 What do you think will happen to an animal suffering from this disease? *(Hint 8)*

Hints and answers follow

Nutrition

1. Think about the major food groups found in each of the meals.

2. There are two different solutions needed for these two tests.

   What temperatures do the two tests have to be carried out at?

3. To do a lot of exercise you need energy. Energy comes from respiration. Which food group is the most important for respiration?

4. Which food group is needed for repair of damaged tissues and healthy teeth and gums?

5. Saliva contains amylase.

   Think about what amylase breaks down, and whether bread or meat are good sources of this.

6. How has the food been changed by the time it reaches this point in the digestive system?

   To perform its function region E must have a large surface area. How is this achieved?

7. The amount of food assimilated depends on the amount of food absorbed at the small intestine. What will happen to the ability of the small intestine to absorb food?

8. What is the effect on an animal of not getting enough of all the food groups?

Answers

1 see table in *Improve your knowledge* **2** starch – iodine test / simple sugar – Benedict's test **3** diet should contain a large amount of carbohydrates to provide the energy needed (through respiration) / lack of vitamin C (scurvy) / eaten citrus fruit **5** A–mouth / B–oesophagus / C–stomach / D–liver / E–small intestine **6** mouth or pancreas / amylase hydrolyses (breaks down) starch to simple sugars / starch is found in bread not meat **7** absorb digested food / long / covered in villi / good blood supply **8** amount of food absorbed will decrease / so less can be assimilated **9** animal will become thin / lack energy and may develop deficiency diseases

19

Breathing and respiration

Test your knowledge

1 Breathing is the process of taking in oxygen and releasing _____ _____ using the _____ .

2 Air passes through the nose or mouth, down the windpipe or _____ , through the bronchi and bronchioles into microscopic air sacs called _____ .

3 Alveoli have thin walls and a good blood supply for rapid _____ .

4 When we inhale, the rib cage moves upwards and _____ and the diaphragm _____ . Exhaled air contains more _____ _____ and it is warmer and _____ than inhaled air.

5 Respiration is the release of _____ from food.

6 Aerobic respiration uses _____ contained in the air we breathe and _____ in the food we eat.

7 The energy released during respiration can be used to help build up large molecules from _____ ones and to help us keep warm.

8 Cigarettes contain many harmful substances, which include tar, carbon monoxide and _____ .

Answers

1 carbon dioxide / lungs **2** trachea / alveoli **3** diffusion **4** outwards / flattens / carbon dioxide / wetter **5** energy **6** oxygen / glucose. **7** smaller **8** nicotine

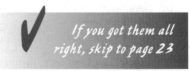

If you got them all right, skip to page 23

Breathing and respiration

Improve your knowledge

1 **Breathing** takes air into and out of the body using the lungs. This provides cells with oxygen and removes carbon dioxide. The lungs are found in the upper part of the body known as the thorax (chest) which is protected by the ribs and separated from the lower part of the body by the diaphragm.

2 Practise labelling the structure of the thorax as below:

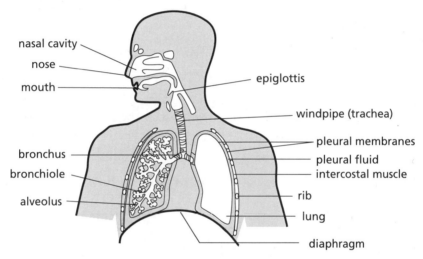

The air passes into our nose or mouth, down the **trachea** (windpipe) and this splits into two **bronchi**, one going to each lung. These then divide into smaller tubes called **bronchioles**, which end in microscopic air sacs called **alveoli**.

The trachea, bronchi and bronchioles have **ciliated** cells which produce mucus. The mucus is sticky to trap bacteria and dirt and is swept by the tiny hairs to the top of the trachea. This helps to prevent infection.

3 **Gaseous exchange** takes place between the alveoli and the blood capillaries. The alveoli are specialised for gaseous exchange because they have very thin walls and have a very good blood supply. There are 350 million alveoli, which makes a very large surface area. This speeds up the diffusion of gases.

Oxygen diffuses from the alveoli into the blood capillary and carbon dioxide diffuses from the blood capillary into the alveoli.

4 Air which enters the lungs is called **inhaled** air and air which leaves the lungs is called **exhaled** air. Compare the differences between inhaled and exhaled air in the table shown below:

	Inhaled air	Exhaled air
Oxygen	21%	16%
Carbon dioxide	0.04%	4%
Water vapour	amount of water in the air	saturated
Nitrogen	79%	79%

5 Respiration is the release of energy from food. Respiration is most efficient when oxygen is available. The equation for aerobic respiration is as follows.

glucose + oxygen → **ENERGY** + carbon dioxide + water

6 The glucose for respiration is contained in the carbohydrates we eat. The oxygen for respiration is contained in the air we breathe in. The carbon dioxide and water produced in respiration are breathed out.

7 The energy which is released during respiration can be used to:
- build tissues
- help muscles contract
- help maintain body temperature in cold conditions.

8 **Smoking** causes damage to the lungs because the smoke contains tar, which stops the tiny hairs working properly. Two other harmful substances are also released from cigarettes – carbon monoxide and nicotine. Carbon monoxide prevents as much oxygen being carried around the blood. Nicotine stimulates nerve impulses and is addictive. Diseases caused by smoking include bronchitis, lung cancer and emphysema. Babies born to women who smoke are often smaller than average and there is an increased chance of premature birth and, in some cases, death.

Now learn how to use this knowledge

Breathing and respiration

Use your knowledge

The diagram below shows the section of the thorax.

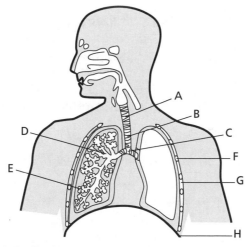

1 Name the parts labelled A to H. *(Hint 1)*

2 Explain how air enters the lungs. *(Hint 2)*

3 Complete the table below to show the differences between inhaled and exhaled air. *(Hint 3)*

	Inhaled air	**Exhaled air**
Oxygen	A _____	16%
Nitrogen	79%	B _____
Carbon dioxide	C _____	4%
Temperature	Air temperature	D _____

4 What process is the oxygen needed for? *(Hint 4)*

5 Explain why, when we breathe on a cold mirror, a layer of condensation forms. *(Hint 5)*

Hints and answers follow

Breathing and respiration

1 Learn this – look at number 2 in *Improve your knowledge*.

2 Learn this – look at number 2 in *Improve your knowledge*.

3 Learn this – look at number 4 in *Improve your knowledge*.

4 Oxygen is needed for the process to release energy.
What is this called?
Look at number 5 in *Improve your knowledge*.

5 What is contained in exhaled air?

Answers

1 A – trachea / B – cartilage / C – bronchus / D – bronchiole / E – alveolus / F – rib / G – intercostal muscle / H – diaphragm **2** muscles and diaphragm contract, this flattens the diaphragm. The volume of the thorax increases and air rushes in **3** A – 16% / B – 79% / C – 0.04% / D – 37°C **4** respiration **5** exhaled air is saturated with water

Blood and circulation

Test your knowledge

10 minutes

1 The blood has three major functions. These are transport, _____ and _____ .

2 Blood is made up of two parts: _____ , a straw-coloured liquid, and _____ , which make up the solid part.

3 The _____ shape of red blood cells means that they have a large _____ _____ to carry _____ . They are _____ than white blood cells and have no _____ .

4 _____ carry blood away from the heart. They have thick _____ walls.

5 The _____ side of the heart pumps blood to the lungs and the left side pumps blood around the _____ .

6 In the organs, blood flows through tiny vessels called _____ .

7 Each side of the heart has two halves. The top halves are called _____ and these pump blood into the thicker-walled _____ which then pump blood to the lungs and body.

8 _____ in the heart ensure that blood flows in the right direction. _____ which carry blood back to the heart also have valves.

Answers

1 defence / clotting **2** plasma / cells **3** biconcave / surface area / oxygen / smaller / nucleus **4** Arteries / elastic **5** right / body **6** capillaries **7** atria / ventricles **8** Valves / Veins

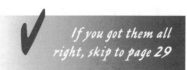
If you got them all right, skip to page 29

25

Blood and circulation

Improve your knowledge

20 minutes

1. The blood has three important functions:
   a) **Transport** of oxygen, food, wastes (carbon dioxide and poisonous chemicals e.g. urea), hormones and heat.
   b) **Fighting infection** against invading organisms (e.g. bacteria and viruses) which cause disease and infection.
   c) **Blood clotting** to stop the loss of blood at wounds.

2. Blood is made up of two parts:
   a) **Plasma** (straw-coloured liquid) – 90% water and 10% dissolved substances
   b) **Cells** (non-liquid part) – red blood cells, white blood cells and platelets.

3. The important thing to understand about blood cells is that their appearance directly relates to their job within the blood:

Cell	Function	Appearance	Explanation of adaptations
Red blood cells	• Transport of oxygen from lungs to respiring cells	• Biconcave • No nucleus • Contain the red pigment haemoglobin	• Large surface area to absorb oxygen • Can pack in more haemoglobin • Haemoglobin carries oxygen
White blood cells	• Fighting disease and infection by attacking organisms invading the body	• Large cells with different shaped nuclei	• Large size: helps cells attack invading organisms
Platelets	• Prevent blood loss during injury	• Fragments of whole cells	• Able to stick together to form a plug • Produce enzymes to clot blood (make solid 'lumps')

4 **Blood vessels** are branched tubes transporting blood to every part of the body. Their appearance varies depending on the job they perform:

Blood vessel	Function	Appearance	Explanation of adaptations
Arteries	• Carry blood away from the heart under high pressure • Mainly carry blood with oxygen in (oxygenated)	• Walls have elastic outer layer and thicker muscle layer	• Elastic and muscle layers: absorb pressure produced by heart • Muscle layer: squeeze blood to help move it along
Capillaries	• Carry blood from arteries to veins very slowly • Allow oxygen and food to go into cells, and waste to be removed into the blood	• Walls are very thin (one cell thick) and have holes (pores) in them	• Thin walls: useful products and wastes only have a short distance to travel • Pores: to allow fluid to leave blood
Veins	• Carry blood back to the heart under low pressure (due to distance from heart) • Mainly carry blood which has lost its oxygen (deoxygenated)	• Walls much *thinner* than arteries (thicker than capillaries) • *Valves* present (flaps on the inside wall)	• Valves: stop blood flowing backwards due to the low pressure

5 The blood is pumped around the body by the **heart**, found between the two lungs. It can be considered as two pumps joined together.

Double circulation of blood through the heart of mammals

The two pumps **contract** (squeeze) at the same time. The blood passes through the heart twice before going back around the body – this is what is meant by a double circulation system.

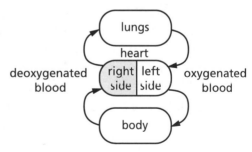

Look at the diagram very carefully – the left- and right-hand side are reversed. The *right side of the heart* is on the left-hand side, and the left side of the heart is on the right-hand side. This may sound a little confusing, but it is because of the sides of the body that they are on.

6 Once the blood reaches the organs it passes from the arteries into very narrow, thin-walled vessels called **capillaries**. Useful substances such as oxygen pass from the blood in the capillaries into the cells. Waste products such as carbon dioxide pass from the cells into the blood in the capillaries to be taken back to the right side of the heart.

7 The heart is divided into two halves. The top halves are thin-walled **atria** (the **right atrium** and the **left atrium**), and the bottom halves are thicker muscular-walled **ventricles** (the **right ventricle** and the **left ventricle**). Blood enters the atria first, which pumps the blood into the ventricles. The ventricles pump the blood out of the heart to the lungs and body. Ventricles have thicker muscular walls than the atria because they pump the blood a greater distance.

Structure of the heart

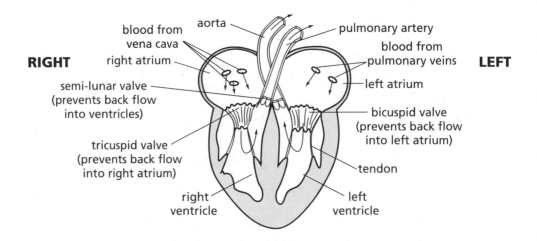

8 **Valves** between the atria and ventricles make sure blood leaves the heart from the ventricles, and does *not* flow back to the atria. The **bicuspid valve** is on the left-hand side, and the **tricuspid valve** on the right-hand side. **Semi-lunar valves** in the aorta and the pulmonary artery stop blood flowing back into the ventricles. Veins carrying blood back to the heart also have valves.

Now learn how to use this knowledge

Blood and circulation

Use your knowledge

20 minutes

These cells are taken from human blood.   cell A   cell B

1 Name the other major component of blood apart from blood cells.

Hint 1

2 State one function of cell A and one function of cell B.

3 Identify and explain **two** visible features of cell A that help it perform its function.

Hint 2

4 During its lifetime cell B is able to leave the blood and move among the body cells. Which type of blood vessel contains pores which allow cell B to do this? Identify **one** other characteristic of this type of blood vessel.

Hint 3

The diagram shows a simplified heart with the main blood vessels.

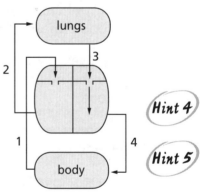

5 Which of the blood vessels contains blood at the highest pressure in the body?

Hint 4

6 What is the major difference between the content of the blood in vessel 3 compared to the blood in vessel 2?

Hint 5

The blood in blood vessel 1 contains blood at very low pressure.

7 Explain what problems this could cause to the circulation of the blood.

Hint 6

8 How does this type of blood vessel stop this problem occurring?

Hint 7

9 Explain one way that the appearance of the left-hand chamber is related to its function.

Hint 8

✓ *Hints and answers follow*

Blood and circulation

1. Cells are the solid component of the blood. What is the name of the liquid component the cells are carried in?

2. Remember to compare structure to function. How is cell A different from a normal cell? These are the features that help it perform its function.

3. Which blood vessels are closest to individual cells?

    The function of these blood vessels is to allow easy exchange of materials such as food and gases between the blood and the cells. What characteristics should the blood vessels have for this function?

4. Think how blood pressure is produced – the blood vessel nearest where the blood pressure is produced must have the highest blood pressure.

5. Blood in vessel 2 is flowing to the lungs, and blood in vessel 3 is flowing away from the lungs. What do the lungs do and how does this change the blood?

6. Blood pressure makes the blood flow in one direction. What may happen to blood under low pressure moving up the legs from the feet?

7. The same principle is used in the heart to stop the blood flowing in the wrong direction.

8. The heart is simply a pump made up of muscle. Think where this chamber pumps the blood to and what this will do to the amount of muscle needed, compared to other chambers of the heart.

Answers

1 liquid plasma 2 A – transport of oxygen / B – defence against infection 3 no nucleus – increases volume available to carry O_2 / biconcave shape – increases surface area to take up O_2 4 capillaries / vessel wall only one cell thick 5 4 – aorta 6 vessel 3 – blood is oxygenated, vessel 2 – blood is deoxygenated 7 blood may flow backwards/not be returned to the heart 8 one-way valves in the wall 9 thick muscular wall to pump the blood away from the heart around the body

Nerves and hormonal coordination

Test your knowledge

1. Living organisms detect changes in their environment called _____ using special cells called _____ .

2. In humans the receptors which detect pressure and temperature changes are found in the _____ . In plants the receptors which detect light are in the _____ .

3. The central nervous system is made up of the brain and the _____ _____ .

4. The _____ controls the size of the pupil, which controls the amount of _____ entering the eye.

5. The receptors which detect light are found on the _____ .

6. _____ is the ability to maintain a constant internal environment.

7. If a person has kidney failure, either a kidney machine can be used or a kidney _____ can take place.

8. If a person gets too hot, the blood vessels near the surface of the skin _____ to allow heat to escape by _____ . The sweat glands produce sweat, which help cool you down by the process of _____ .

Answers

1 stimuli / receptors 2 skin / leaves 3 spinal cord 4 iris / light 5 retina 6 Homeostasis 7 transplant 8 get bigger/dilate / radiation / evaporation

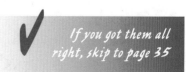

If you got them all right, skip to page 35

Nerves and hormonal coordination

Improve your knowledge

1. Living organisms detect changes in their environment called **stimuli**, e.g. taste, touch, smell, light, sound and balance. **Receptors** are special cells that detect stimuli.

2. Both plants and animals have receptors. Learn the following list:

Animals		Plants	
Stimuli	*Site of receptors*	*Stimuli*	*Site of receptors*
Light	Eyes	Light	Leaves
Sound	Ears	Moisture	Leaves
Change in position	Ears	Gravity	Roots
Taste and smell	Tongue and nose		
Pressure and temperature changes	Skin		

By detecting these stimuli, plants can grow towards areas where there is most light and moisture, to increase their chances of survival.

3. The nervous system has a control centre called the **central nervous system**, which is made up of the **brain** and the **spinal cord**. Nerves carry messages to and from the central nervous system.

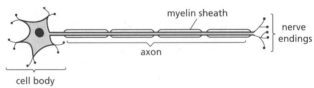

4. Learn the structure and function of the eye:

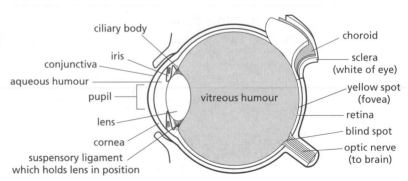

cornea: clear window at the front of the eye. This is the surface where the greatest amount of refraction (bending) of light takes place.

iris: coloured part. This controls how much light can enter the eye.

pupil: black hole in the middle of the iris. This is made bigger or smaller by the muscles in the iris and allows light to enter the eye.

lens: focuses light onto the back of the retina and refraction occurs.

suspensory ligaments: hold the lens in place.

ciliary body: contains muscle which changes the shape of the lens.

sclera: outer tough, white protective layer.

retina: contains rod cells which detect black and white and cone cells which detect colour.

optic nerve: carries messages to the brain. The image is upside down and the brain learns to turn it up the right way.

5. Light from an object enters the eye through the cornea. The curved cornea along with the lens produce an image on the retina. This stimulates receptors which send an electrical impulse to the brain. The brain re-creates the image.

6. **Homeostasis** is the ability to maintain a constant internal environment. Blood sugar level, temperature and water are three examples of factors that need to be controlled. Homeostasis involves the nervous system and hormones – chemical messengers made in the endocrine glands. Hormones are released directly into the blood stream.

   The kidneys control the amount of water in the blood and excrete poisonous waste substances (e.g. **urea**). Blood enters the kidneys through the **renal artery** and goes out through the **renal vein**. The kidneys act like complex filters and clean the blood by removing all of the urea and some of the salts and water in a healthy person. This is called **urine**. The **ureter** carries urine from the kidney to the **bladder**, which stores urine. Urine leaves the bladder through a tube called the **urethra** when the muscle relaxes.

7. If a person is suffering from **diabetes**, glucose will appear in the urine and can be tested using a clinistix. If a person is suffering from **kidney damage**, protein will appear in their urine and can be tested using an albustix.

If a kidney stops working completely, a person has **kidney failure**. There are two ways to deal with this problem:
a) **Kidney machines** – this process is called **dialysis**. It acts like an artificial kidney by filtering out the poisonous waste from the blood. A patient must visit the hospital three times a week for 10 hours.
b) **Kidney transplant** – a healthy kidney from a **donor** is transplanted into the patient's body. However, the new kidney may be rejected by the patient and a further transplant may be needed.

8 The regulation of temperature in the body is called **thermoregulation**. The skin is the organ which regulates body temperature. Learn how to label the skin:

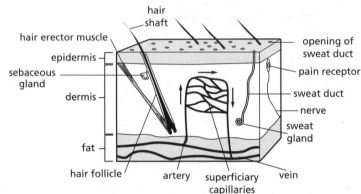

If a person gets too hot:
a) The sweat glands produce sweat, which contains water. This cools the skin down because the heat from the skin changes the water into steam by **evaporation**.
b) The blood vessels near the surface of the skin get bigger. This allows heat from the blood to escape by **radiation**.
c) The hair erector muscles relax so that the hairs lie flat against the skin.

If a person gets too cold:
a) The sweat glands do not produce sweat.
b) The blood vessels near the surface of the skin get smaller. So less heat escapes.
c) The hair erector muscle contracts so that the hair stands up. This traps a layer of air around the skin, which keeps us warm by **insulation**. In humans, this is called goose pimples!
d) The muscles contract in a process called **shivering**, which keeps us warm.
e) There is a fatty layer in the skin, which keeps us warm (**insulation**). This is a very thick layer in animals like polar bears that live in the Arctic.

✓ *Now learn how to use this knowledge*

Nerves and hormonal coordination

Use your knowledge

1 The pupils in our eyes show a reflex action to light. Draw in the pupils in the diagrams below to show the appearance of the eyes.

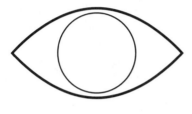   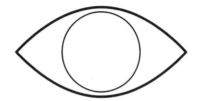

      dim light   bright light

2 Urine can be tested to see if a person has diabetes or kidney damage. If they have diabetes, sugar will be present. If they have kidney damage, protein will be present. The urine of a normal person will contain water, salts and urea. The urine of three students was tested in a random drugs test. The results are shown below.

Name	Contents of urine
Michael	water, salts, urea and sugar
Emma	water, salt and urea
Maria	water, salt, urea and protein

a) Which person is suffering from diabetes?   (Hint 2)

b) Which person is suffering from kidney damage?   (Hint 3)

Hints and answers follow

Nerves and hormonal coordination

1. The pupils get bigger and smaller in different lights. Draw the correct appearance of the pupil.

2. Which person has sugar in their urine?

3. Which person has protein in their urine?

Answers

1 dim light – large pupil; bright light – small pupil 2 a) Michael b) Maria.

Genetics

Test your knowledge

1. No two individual organisms on Earth look the same. These differences in characteristics are known as _____ . Baby rabbits look like their parents because of information passed on in the form of _____ . These are found on chromosomes in the _____ of each cell.

2. In body cells the chromosomes are normally found in _____ . The genes on the chromosomes exist in different forms called _____ .

3. Gametes have _____ the number of chromosomes which body cells have. They are therefore called _____ . Cells with the full number of chromosomes are said to be _____ .

4. In _____ reproduction, two gametes fuse together. In asexual reproduction there is no fusion of gametes and only _____ individual is needed. Sexual reproduction produces _____ variation.

5. In humans the female sex chromosomes are _____ whilst the male sex chromosomes are XY. These can be used to show that the probability of a baby boy or girl is _____ .

6. Some alleles are _____ . If they are inherited they may cause disease.

7. By breeding selected individuals together we can _____ useful characteristics in offspring.

8. If the environment changes species may become _____ , i.e. die out. Dead organisms may be preserved as _____ . By looking at the fossil record we can observe gradual changes in the characteristics of organisms over millions of years. Fossils therefore produce evidence of _____ .

Answers

1 variation / genes / nucleus 2 pairs / alleles 3 half / haploid / diploid 4 sexual / one / greater 5 XX / 50% 6 harmful 7 combine 8 extinct / fossils / evolution

If you got them all right, skip to page 41

37

Genetics

Improve your knowledge

1. Every organism on Earth looks different from every other organism, even individuals of the same species are different. These differences in characteristics are known as **variation**.

   Plant and animal characteristics are controlled by two factors:
   a) **Genes** which are inherited from parents. This is why young plants and animals resemble their parents. Genes are found in the nucleus on **chromosomes**; each chromosome carries many genes which control different characteristics.
   b) **Environment**. For example, imagine two plants with identical genes. If one of the plants is sheltered from wind, it will grow taller than a plant growing in an exposed, windy environment.

2. In body cells the chromosomes are found in pairs. Each of the pair contains the same number and type of genes at the same positions. Genes can exist in different forms called **alleles**.

3. Chromosomes are present as pairs in all body cells with nuclei (i.e. they are absent in red blood cells). Cells with the full number of paired chromosomes are said to be **diploid**. In humans the **diploid number is 46** (23 pairs of chromosomes).

   Each pair of chromosomes is made up of one from the male parent and one from the female parent, formed when the sperm and the egg cells join in fertilisation. This means that gametes are haploid, having half the full number of chromosomes, with no pairs. In humans, the **haploid number is 23**. By halving the number of chromosomes the diploid number stays the same in every generation (see diagram).

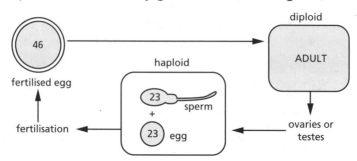

4 There are two types of reproduction. In sexual reproduction, two haploid gametes, one from the male and one from the female fuse together. This gives great variation. In asexual reproduction there is no fusion of gametes and only one individual is needed. When we take cuttings from plants and grow them into full specimens we are carrying out asexual reproduction. The offspring are identical to the parent, i.e. there is no variation.

5 One of the 23 pairs of chromosomes are called **sex chromosomes**, because they control whether a fertilised egg becomes a male or female. These sex chromosomes are described by their appearance and can be either:

a) **X chromosome** (long chromosome) or

b) **Y chromosome** (short chromosome).

We can use genetic diagrams to show there is an equal chance of a baby being a boy or a girl:

Phenotype:	Male				Female	
Genotype:	XY				XX	
Possible gametes:	X	Y	fertilisation		X	X
Children's genotype:	XX	XX			XY	XY
Children's phenotype:	Female	Female			Male	Male
			therefore 50% male and 50% female.			

6 Some genes produce harmful effects when inherited. For example, **Huntington's chorea** – a disease which affects the nervous system – can be inherited even if just one of the parents has the disease. Cystic fibrosis, on the other hand, can only be inherited if both parents have the gene for the disease.

7 **Selective breeding** or **artificial selection** is when humans choose the organisms which they allow to breed together, e.g. Jersey cows with high milk yields are bred together. Offspring produced are better for farming because they inherit these useful characteristics, and the high yields make more money.

Cross-breeding is the crossing of different varieties of plants and animals to produce offspring with features from both varieties. For example, imagine that we have two varieties of rice: one is high-yielding but has such a long stem that it keeps blowing over and the other has a short stem but lower yield. By cross-breeding we can obtain a short-stemmed high-yielding variety.

8 If the environment changes, some animals and plants may die out. **Fossils** are the remains of animals or plants, some of which died (or became extinct) millions of years ago. Usually, dead organisms are broken down by bacteria and fungi (i.e. they decay) but sometimes parts of animals may become trapped in amber. Inside the amber there is no oxygen so decay is impossible. The organism is preserved as a fossil.

By looking at all of the fossils which have been found for a particular organism and by calculating when the fossil was formed, we can learn how the organism has changed or evolved over millions of years.

Even if the environment changes an organism may not die out. Accidental changes in the genes may give them new characteristics which allow them to survive in the new environment. Such changes, over millions of years, are known as **evolution**.

Now learn how to use this knowledge

Genetics

Use your knowledge

	Male	Female
Parents' body cells:	☐	☐
Gamete cells:	☐	☐
	sperm	eggs
Child's body cells:	☐	

1 In spaces in the diagram above, fill in the number of chromosomes for each of the body cells. *(Hint 1)*

2 Body cells are said to be diploid for chromosome number. What is the name given to the number of chromosomes in sperm and egg cells? *(Hint 2)*

3 Explain the significance of the difference in the number of chromosomes in body cells compared to gamete cells. *(Hint 3)*

✓ *Hints and answers follow*

Genetics

Hints

1. How are body cells and gametes different in the number of chromosomes that they contain?

2. Diploid cells contain a full set of chromosomes. Sperm and egg cells have half the number of chromosomes and have a different word to describe them.

3. If the sperm and egg cells had the same number of chromosomes as the body cells, when fertilisation occurred what would happen to the number of chromosomes in the child compared to the parent?

Answers

1 parents' body cells 46 / gametes 23 / child's body cells 46 2 haploid 3 to avoid doubling of chromosome number in each generation (fertilisation between two haploid gametes restores diploid number)

Ecology
Test your knowledge

1. The ultimate source of energy on the Earth is the _____ .

2. In a food chain the direction of the arrows shows the transfer of _____ . A _____ _____ is a series of interconnected food chains. The pyramid of numbers shows the number of organisms at each _____ _____ or feeding level. The pyramid of biomass shows the _____ of organisms at each trophic level. Biomass decreases at each level because _____ is lost at each level. The greatest biomass is therefore found at trophic level _____ .

3. Decomposition is the breakdown of dead plant and animal material by organisms such as _____ and _____ . Decomposition is fastest in _____ and wet conditions.

4. In the carbon cycle, CO_2 is released into the atmosphere during plant and animal _____ . During photosynthesis, CO_2 is converted into _____ such as glucose. Herbivores obtain their carbon compounds by _____ plants.

5. The place where a community lives is called its _____ .

6. Plants compete with each other for _____ , light and nutrients. Animals may compete with each other for _____ , mates and territory. Competition leads to the survival of _____ _____ .

7. As human population increases more _____ fuels are burned to provide _____ . Burning coal in power stations releases _____ _____ and _____ _____ . Nitrogen dioxide is released from _____ exhausts. All of these gases dissolve in rainwater to form _____ _____ .

Answers

1 Sun 2 energy / food web / trophic level / mass / energy / one 3 bacteria / fungi / hot 4 respiration / carbohydrates / eating 5 habitat 6 water / food / the fittest 7 fossil / energy / carbon dioxide / sulphur dioxide / vehicle / acid rain

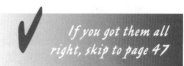

If you got them all right, skip to page 47

Ecology

Improve your knowledge

1 Solar energy is converted into chemical energy by green plants during photosynthesis. These green plants are known as primary producers. Primary producers are eaten by herbivores or primary consumers, which in turn may be eaten by carnivores (secondary consumers). Feeding relationships can be shown in food chains and webs.

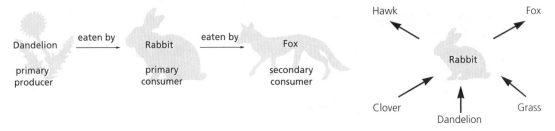

2 The organisms in the **food chain** or **web** can be organised into different feeding or **trophic levels**.

The total number of organisms at each trophic level can be shown in a pyramid of numbers while a pyramid of biomass shows the mass of all the organisms at each trophic level.

Secondary consumers = Trophic level 3

Primary consumers = Trophic level 2

Primary producers = Trophic level 1

Pyramid of numbers based on an oak tree

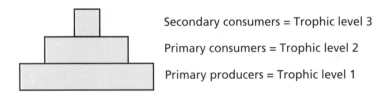

10 blue tits

5000 caterpillars   mass decreases

1 tree

Biomass decreases at each trophic level because energy is lost, often as heat, at each level.

3 **Decomposition** is the breakdown of dead organisms by bacteria and fungi. It is fastest in hot and wet conditions. During decomposition many nutrients escape into the soil, where they can be recycled by plant roots.

Carbon cycle

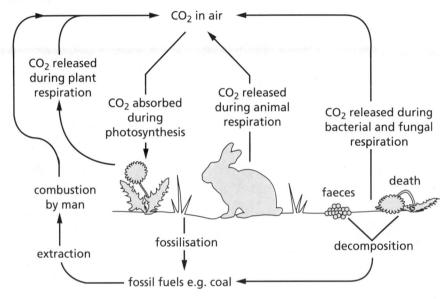

4 Carbon is found in every living organism. Only green plants can change CO_2 into sugars which plants and animals can then use. Respiration and decomposition release CO_2 back into the atmosphere. This is the **carbon cycle**.

5 You must learn the following definitions:

Population All the members of a species, e.g. rabbits living in a particular area at the same time

Community All the different populations of organisms which live together – plants, dandelions, rabbits, foxes, etc.

Habitat The place where a community lives, e.g. a meadow

Environment The conditions which exist in a habitat

Adaptation A feature of an organism which increases its chances of survival in a particular environment, e.g. camouflage.

6 Animals of the same species and those of different species compete with each other for food and territories. Plants compete with each other for nutrients, water and light. Competition means that there are winners and losers. Within a fox population, there will be competition for food such as rabbits. The fastest foxes will catch many more rabbits than the slowest fox which may then starve to death. Competition leads to the survival of the fittest.

7 The human population explosion has caused severe air and water pollution:

Substance	Source	Effect
Sulphur dioxide	Power stations burning fossil fuels	Dissolves in moisture to form acid rain
Nitrogen dioxide	Vehicle exhaust fumes	Dissolves in moisture to form acid rain
Carbon monoxide	Incomplete combustion of petrol	Odourless, invisible, poisonous gas
CFCs (chlorofluorocarbons)	Aerosols, refrigeration, foam products	Destroy ozone layer, increasing chances of skin cancer
Sewage	Sewage plant floods, e.g. in heavy rainfall	May contain metals, e.g. copper. Will contain pathogenic (disease-causing) bacteria
Nitrates	Dissolved nitrate fertilisers which drain from agricultural land	Lead to algal blooms which are quickly broken down by bacteria. These bacteria use up oxygen in the water, killing fish, etc.
Phosphates	Erosion from agricultural land	As with nitrates, phosphates encourage algal blooms

Now learn how to use this knowledge

Ecology

Use your knowledge

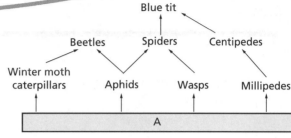

The diagram shows a food web in a woodland.

1 Name the source of energy in the food web. *(Hint 1)*

2 What type of organisms would be in box A? *(Hint 2)*

3 Name a secondary consumer in the food web. *(Hint 3)*

4 Which trophic level would have the least biomass? Explain your answer. *(Hint 4)*

5 Explain how carbon atoms in the feathers of the blue tit could eventually end up as carbon dioxide in the atmosphere. *(Hint 5)*

The graph shows the population of mice and goshawks in a coniferous wood. The goshawks, which eat mice, rabbits and small birds, first arrived in the wood in 1990. The mice eat seeds and buds.

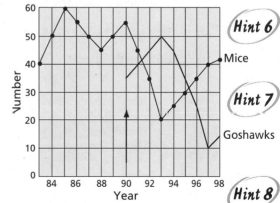

6 In this example, which species is the predator and which is the prey? *(Hint 6)*

7 Suggest reasons why the mouse population went up and down before the goshawks arrived in the wood. *(Hint 7)*

8 Suggest an explanation for the change in the population of the organisms between 1990 and 1997. *(Hint 8)*

Hints and answers follow

Ecology

Hints

1. Learn this!

2. These organisms convert solar energy into the chemical energy which herbivores can use.

3. Learn this.

4. What happens to the amount of energy available at each trophic level?

   What happens to the energy which is not passed up the food web?

5. What will eventually happen to the blue tit?

   What kind of organisms break down dead material?

   What process do all living things carry out?

6. Learn the definitions.

7. What factors influence the size of the population?

8. What happened to the goshawk population in 1990–2?

   What happened to the mouse population over that period?

   Think 'predator-prey'. How does one population affect the other?

Answers

1 Sun 2 primary producer 3 beetle, spider or centipede 4 top or trophic Level 4 / because energy is lost between each stage 5 Blue tit dies and is decomposed. Carbon atoms are released as CO_2 into atmosphere 6 predator = goshawk / prey = mice 7 food supply / disease / changes in weather 8 Initially there is plenty of food for goshawks, so their population increases. Then the mice population decreases because many are eaten. Goshawks then starve and their population declines. Mice population recovers.

The periodic table
Test your knowledge

1 The elements in the periodic table are arranged in horizontal rows called _____ and vertical columns called _____ .

2 Most elements are metals. They usually have _____ melting and boiling points, are good conductors of _____ and _____ and are usually physically _____ . The other elements are called _____ . They have _____ melting and boiling points and are usually physically _____ .

3 Elements are placed in order of increasing _____ _____ . As we go across a period we are filling up the _____ _____ with electrons.

4 Elements in the same group have similar _____ _____ because they have the same _____ structure.

5 Going down the group there is a _____ change in physical and chemical properties.

Answers

1 periods / groups. **2** high / heat / electricity / strong / non-metals / low / weak **3** atomic number / outer shell **4** chemical properties / electronic **5** gradual

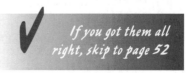

If you got them all right, skip to page 52

The periodic table

Improve your knowledge

1 Part of the periodic table is show below.

1	2		key			3	4	5	6	7	0
H			☐ metal								He
Li	Be					B	C	N	O	F	Ne
Na	Mg		▨ non-metal			Al	Si	P	S	Cl	Ar
K	Ca									Br	Kr
Rb			Transition metals e.g. Fe, Cu, Zn							I	Xe
Cs											Ra

The elements are arranged in horizontal rows called **periods** and also vertical columns called **groups**, numbered from 1 to 0 as shown (the 8th group is called group 0).

2 More than three-quarters of the elements are **metals**, the rest are called **non-metals** and are found in the top right-hand corner. The metals in the middle of the periodic table are called **transition metals**. Properties of metals and non-metals are shown below.

Metals	**Non-metals**
Solid at room temperature (except mercury)	Many are gases
Most have high melting and boiling points	Most have low melting and boiling points
Shiny when freshly cut or scratched	Usually dull and coloured
Good conductors of electricity and heat	Poor conductors of heat and electricity
Mostly strong, but easily bent into shapes	Weak and brittle
Form ionic bonds with non-metals	Form ionic bonds with metals and covalent bonds with other non-metals
Oxides are basic	Oxides are acidic

3 The elements are placed in order of **increasing atomic number** as read from left to right. As we go across a period an extra proton is added to the nucleus and an extra electron in the outer shell. Going from one end of a period to the other corresponds to filling the outer shell with electrons, e.g. from sodium, Na **2,8,1** to argon, Ar **2,8,8**.

4 Elements in the same group have the same outer-shell electronic structure and so have similar chemical properties. Group 1 elements are all metals which form +1 ions and group 7 elements are all non-metals which form −1 ions.

5 There is a gradual trend in the properties of the elements as we go down the group. Going down group 1 melting points decrease but reactivity increases. Going down group 7 melting points increase but reactivity decreases.

Now learn how to use this knowledge

The periodic table

Use your knowledge

Francium, Fr, comes at the bottom of group 1. This element is radioactive and is very difficult to isolate. However chemists can predict some of its properties by looking at the properties of the other elements in group 1 and the trends down the group.

1. Do you think francium would be a metal or non-metal? *(Hint 1)*

2. How many electrons would it have in its outer shell? *(Hint 2)*

3. What kind of bonding do you think is present in francium chloride? *(Hint 3)*

4. Write the formula you would expect for this compound. *(Hint 4)*

Potassium and lithium are elements in group 1. Potassium melts at 64°C and lithium melts at 180°C.

5. Compared to these numbers, at what temperature do you think francium might melt? *(Hint 5)*

Lithium and potassium react with cold water to form hydrogen and a solution of the metal hydroxide. Lithium reacts quite slowly but potassium reacts very quickly and gives out a lot of heat.

6. Do you think francium would be more or less reactive than potassium? *(Hint 6)*

7. Write a word equation for the reaction between francium and water. *(Hint 7)*

A chemist investigates two unknown elements. Both are solids but one is a metal and one is a non-metal.

8. The chemist cut the elements. How might you tell them apart by their appearance? *(Hint 8)*

✓ *Hints and answers follow*

The periodic table

1. What are the other elements in the group?
2. What group is it in?
3. The compound contains metal and a non-metal.
4. What groups are the two elements in and what are their valencies?
5. What is the trend **down** the group?
6. Is potassium more or less reactive than lithium? What is the trend?
7. What are the reactants and products?
8. What do metals look like when freshly cut?

Answers

1 metal 2 1 3 ionic 4 FrCl 5 lower than 64°C 6 more reactive 7 francium + water → francium hydroxide and hydrogen 8 metal would be shiny, non-metal wouldn't

Chemical equations

Test your knowledge

1 Write the formula and relative formula mass of the following compounds. The number of each atom present is given in brackets.

   a) Hydrogen (1) chloride (1) _____

   b) Calcium (1) chloride (2) _____

2 Write the formula of the following compounds.

   sodium oxide _____   calcium iodide _____   hydrogen bromide _____

3 Complete the following word equation.

   zinc + _____ → zinc sulphate + copper

4 Write the following word equation in symbols.

   sodium + water → sodium hydroxide + hydrogen
   _____

5 Balance the following equation.

   $C_3H_8 + \underline{5}\,O_2 \rightarrow \underline{3}\,CO_2 + \underline{4}\,H_2O$

6 Write a balanced equation for the reactions between nitrogen + hydrogen to form ammonia _____

7 Fill in the missing state symbols for the following equation.

   $CaCO_3(s) + H_2SO_4(aq) \rightarrow CaSO_4\ \underline{\quad} + CO_2\ \underline{\quad} + H_2O\ \underline{\quad}$

Answers

1 a) HCl/36.5 b) $CaCl_2$/111   2 Na_2O / CaO / HBr   3 copper sulphate   4 $2Na + 2H_2O \rightarrow 2NaOH + H_2$ 5 $C_3H_8 + 5O_2 \rightarrow 3CO_2 + 4H_2O$ 6 $N_2 + 3H_2 \rightarrow 2NH_3$ 7 $CaSO_4(aq) / CO_2(g) / H_2O(l)$

If you got them all right, skip to page 58

Chemical equations

Improve your knowledge

20 minutes

1. Atoms are represented by the **symbols** shown on the periodic table. Molecules made up of atoms can be represented by using these symbols to create **formulae**.

Compound	Contains	Formula	Rfm
Water	H, O	H_2O	(2 x 1) + 16 = **18**
Sodium chloride	Na, Cl	NaCl	23 + 35.5 = **58.5**
Sodium hydroxide	Na, H, O	NaOH	23 + 16 + 1 = **40**

   The relative formula mass (rfm) is worked out by adding together the relative atomic masses of all the atoms present.

2. **Valency** is how many bonds (ionic or covalent) an atom usually forms. This depends on which group in the periodic table the element is in.

Group	1	2	3	4	5	6	7
Valency	1	2	3	4	3	2	1

   In ionic substances the valency is equal to the charge of the ion.

   When two atoms or ions (A and B) bond, the formula of the molecule formed can be worked out by making sure that

   (number of atom A x valency A) = (number of atom B x valency B)

NaCl	**CaCl$_2$**	**CH$_4$**	**Al$_2$O$_3$**
1 x 1 = 1 x 1	1 x 2 = 2 x 1	1 x 4 = 4 x 1	2 x 3 = 3 x 2

3. A **chemical reaction** is when substances (**reactants**) combine together to form new substances (**products**). Chemical bonds are **broken** in the reactants and **formed** in the products. Chemical reactions are represented by **chemical equations** which tell us which reactants and products are involved. Chemical equations can be written as **word equations**.

   sodium hydroxide + hydrochloric acid → sodium chloride + water
   zinc + sulphuric acid → zinc sulphate + hydrogen
   calcium carbonate + sulphuric acid → calcium sulphate + carbon dioxide + water

55

4 Chemical equations can be written by using symbols and formulae.

$$Zn + H_2SO_4 \rightarrow ZnSO_4 + H_2$$

5 Chemical equations must be **balanced**.

THIS MEANS THAT THE NUMBER OF EACH TYPE OF ATOM ON THE LEFT-HAND SIDE OF THE EQUATION MUST BE THE SAME AS THE NUMBER OF EACH TYPE OF ATOM ON THE RIGHT-HAND SIDE OF THE EQUATION.

H Atoms		2				2
Na Atoms	2			2		
	Na_2CO_3	+ 2HCl	$\rightarrow$	2NaCl	+ CO_2	+ H_2O
C Atoms	1				1	
Cl Atoms		2		2		
O Atoms	3				2	1

6 To balance an equation do the following:

a) Write down the equation using symbols and formulae.

b) Count the number of each type of atom on both sides of the equation.

c) If the numbers are not the same on both sides put the correct number **in front of** the appropriate symbol or formula.

For the reaction between hydrogen, H_2, and nitrogen, N_2, to form ammonia, NH_3:

a) $N_2 + H_2 \rightarrow NH_3$

b) On the left-hand side there are 2 N atoms, on the right-hand side there is only 1.

On the left-hand side there are 2 H atoms, on the right-hand side there are 3.

c) So we must have 2 NH_3 molecules and 3 H_2 molecules. Thus we get the equation

$$\begin{array}{ccc} 2N \quad 6H & & 2N + 6H \\ N_2 + 3H_2 & \rightarrow & 2NH_3 \end{array}$$

To balance an equation start with **one type** of atom and count up the number on the **left-hand side** and make sure it's the same as the number of that atom on the **right side**. If it isn't then you have to put numbers **in front** of the formulae containing that atom to make them the same. Do this for each type of atom in turn.

The equation below has been balanced, the steps show you how.

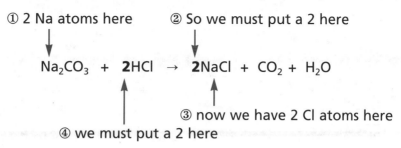

Now, there is the same number of each type on the left-hand side and the right-hand side.

Example: Balance the following equation.

a) Count up the C and H atoms on both sides

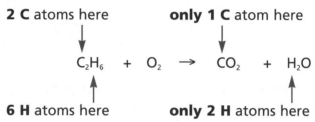

b) Balance these atoms by putting numbers in front of the appropriate formulae

$$C_2H_6 + O_2 \rightarrow 2CO_2 + 3H_2O$$

c) Now add up the O atoms

    2 O atoms here   7 O atoms in total

d) Therefore we must have (at least) 7 O atoms on the left-hand side. However, each molecule of oxygen contains two atoms. So double the number of oxygen atoms on the right-hand side to 14, which produces the final equation:

$$2C_2H_6 + 7O_2 \rightarrow 4CO_2 + 6H_2O$$

7 The physical states of substances can be shown by using the state symbols as follows:

solid – (s); **liquid** – (l); **gas** – (g) and **aq**ueous, for substances dissolved in water – (aq).

Now learn how to use this knowledge

Chemical equations

Use your knowledge

 20 minutes

Phosphorus is in group 5 and reacts with chlorine to form compounds that contain only phosphorus and chlorine.

1 Write down the formula of the product you would expect when phosphorus reacts with chlorine. *(Hint 1)*

2 Write a word equation for the formation of this chloride from its elements. *(Hint 2)*

3 Phosphorus can also have a valency of 5. Write down the formula of the chloride you would expect when phosphorus has a valency of 5. *(Hint 3)*

Magnesium sulphate, $MgSO_4$ is an ionic solid.

4 Calculate the rfm of pure magnesium sulphate. *(Hint 4)*

Solid magnesium sulphate is often contaminated with water trapped in the crystals. This water is not chemically bonded but can be written in the formula as $MgSO_4.xH_2O$, where x represents the number of water molecules. This is called hydrated magnesium sulphate.

5 If the rfm of hydrated magnesium sulphate is 246, calculate the mass of water present in the formula. *(Hint 5)*

Aluminium reacts with oxygen in the air to form aluminium oxide, an ionic solid.

6 What is the formula of aluminium oxide?

(Hint 6)

7 Write a balanced equation for the formation of aluminium oxide.

(Hint 7)

8 When this oxide is dissolved in dilute sulphuric acid it produces aqueous aluminium sulphate solution. If aqueous sodium hydroxide is added a precipitate of aluminium hydroxide is formed as one of the products. Write a balanced equation for the second reaction, showing state symbols.

(Hint 8)

Lead forms two chlorides, $PbCl_4$, a liquid, and $PbCl_2$, a solid. When $PbCl_4$ is heated it forms $PbCl_2$ as one of the products.

9 Write down a balanced equation for this reaction, showing state symbols.

(Hint 9)

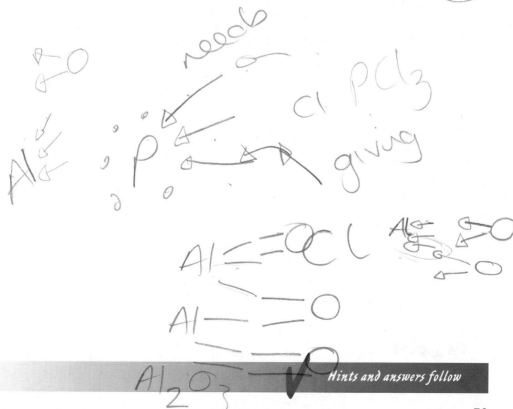

Hints and answers follow

Chemical equations

Hints

1. What is the normal valency for group 5 and 7 atoms?

2. What are the reactants and products of this reaction?

3. How many chlorine atoms will phosphorus bond to if it has a valency of 5?

4. Rfm can be calculated from relative atomic masses on the periodic table.

5. The rfm of hydrated magnesium sulphate is the sum of the rfm of $MgSO_4$ and the H_2O present.

6. Work out the formula from the charges on the ions. What group are these two in?

7. Write down the reactants and products and then balance the equation.

8. What is the formula of aluminium hydroxide and what is its physical state?

9. What must the other product be?

Answers

1 PCl_3 2 phosphorus + chlorine → phosphorus chloride 3 PCl_5 4 120 5 126 6 Al_2O_3
7 $2Al + 1\frac{1}{2}O_2 \rightarrow Al_2O_3$ 8 $Al_2(SO_4)_3(aq) + 6NaOH(aq) \rightarrow 2Al(OH)_3(s) + 3Na_2SO_4(aq)$
9 $PCl_5(l) \rightarrow PbCl_3(s) + Cl_2(g)$

Atomic structure and bonding

Test your knowledge

10 minutes

1 Solids are dense because their particles are _____ _____ . Particles in liquids are able to move, so liquids can _____ .

2 A gas at one end of a tube moves to the other end by a process called _____ .

3 Atoms consist of a central nucleus which contains _____ and _____ . Surrounding the nucleus are _____ contained in _____ .

4 a) The number of protons in an atom is called its _____ _____ . The total number of _____ and _____ is called its mass number. Atoms of an element with different mass numbers but with the same number of protons are called _____ .

b) An atom of aluminium can be written as $^{27}_{13}$Al. Work out how many protons, neutrons and electrons it has.

5 The first electron shell can take ____ electrons and the next two can each take ____ electrons.

6 Atoms form enough bonds to make their outer shell _____ .

7 Ionic bonds are formed when electrons are _____ between atoms. The atom that gains electrons forms a _____ ion, and the one that loses them forms a _____ ion. Metals always form _____ ions.

8 Ionic substances have _____ melting and boiling points, because they have _____ structures. They can _____ _____ when melted. Most covalent substances have _____ melting points.

Answers

1 closely packed / flow/change shape **2** diffusion **3** protons / neutrons / electrons / shells **4** a) atomic number / protons / neutrons / isotopes b) 13 protons / 14 neutrons / 13 electrons **5** 2 / 8 **6** full **7** transferred / negative / positive / positive **8** high / giant / conduct electricity / low

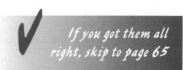

If you got them all right, skip to page 65

Atomic structure and bonding

Improve your knowledge

20 minutes

1 The diagrams show how particles are arranged in solids, liquids and gases.

solid   liquid   gas

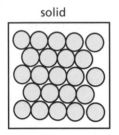 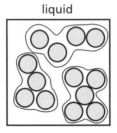 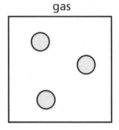

The table shows the properties of solids, liquids and gases:

	Property	**Explanation**
Solid	Dense and can't be compressed Fixed volume and shape	Particles very close together Particles can't move – held together by strong forces
Liquid	Dense and can't be compressed Fixed volume, no shape	Particles close together Particles can flow past each other
Gas	Compressible No shape – expands to fill container	Particles very far apart Particles move freely

2 Diffusion is the process by which gases move and mix. Liquids mix in this way too, but much more slowly.

Diffusion happens because the particles are in constant motion. When particles collide with each other and the sides of the container, they change direction. You can show diffusion is happening by using a coloured gas – e.g. bromine vapour.

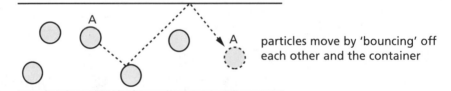

particles move by 'bouncing' off each other and the container

The **heavier** a gas, the **slower** it diffuses.

3 Atoms contain 3 kinds of particles – protons and neutrons in the **nucleus** and electrons in **shells** around the nucleus.

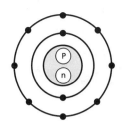

The table shows the properties of protons, neutrons and electrons:

Particle	Mass	Charge
Proton	1	+1
Neutron	1	0
Electron	0	–1

4 Learn the following definitions:

Atomic number = number of protons
Atomic mass = protons + neutrons

In an atom, the number of **electrons** is the same as the number of **protons**.

All atoms of the same element have the same atomic number. Atoms of the same element may sometimes have different atomic mass numbers – such atoms are called **isotopes**.

Atoms are sometimes written like this: $^{16}_{8}O$.

The **top** number – or bigger number – is the **atomic mass**.
The **bottom** number – or smaller number – is the **atomic number**.
The **symbol** tells you what sort of atom it is.

So this atom is **oxygen**, with an **atomic mass** of 16 and an **atomic number** of 8. So we know it has **8 protons, 8 neutrons** and **8 electrons**.

5 The electron shell **closest** to the nucleus can hold **2** electrons. All the others can hold **8**.

You need to be able to work out how electrons are arranged in shells;
e.g. $^{24}_{12}Mg$: the atomic number is 12, so there are 12 protons, and hence 12 electrons.

2 electrons can go in the first shell; another **8** can go in the second shell. We have 2 left, so **2** go in the third shell.

So the **electronic structure** is 2,8,2.

N.B. The **group** of an element in the **periodic table** tells you how many electrons it has in its **outer shell**.

6 Atoms combine by using electrons to make **bonds**. They form enough bonds to make their **outer electron shell full**.

7 **Ionic** bonds are formed by **transferring** electrons from one atom to another.

Metals lose **all** the electrons in their **outer shell** and become positive ions. The **number** of electrons lost gives the **size** of the charge on the ion.

Non-metals **gain** electrons and become **negative ions**. They gain enough electrons to get 8 electrons in their outer shell. The **number** of electrons gained gives the **size** of the charge.

Metals		Non-metals	
Sodium ion	Na^+	Chloride ion	Cl^-
Potassium ion	K^+	Bromide ion	Br^-
Magnesium ion	Mg^{2+}	Iodide ion	I^-
Calcium ion	Ca^{2+}	Oxide ion	O^{2-}
Aluminium ion	Al^{3+}	Sulphide ion	S^{2-}

Ionic bonding always needs to have a **positive** ion and a **negative** ion. So we expect it to happen between **metals** and **non-metals**.

8 Learn these properties of ionic and covalent substances:

Type	Name of structure	Property	Explanation
Ionic	Giant structure	• High melting and boiling point • Conduct electricity when in solution or melted	• Ions strongly bonded. • Ions can move when dissolved or melted
Covalent (most)	Simple molecules	• Low melting and boiling point • Do not conduct electricity	• Weak forces between separate molecules • No charged particles present
Covalent (graphite)	Giant structure	• Very high melting and boiling point • Conducts electricity	• Atoms strongly bonded • Although no ions are present, the bonding allows some electrons to move around
Covalent (diamond, sand)	Giant structure	• Very high melting and boiling point • Do not conduct electricity	• Atoms strongly bonded • No charged particles present

Now learn how to use this knowledge

Atomic structure and bonding

An unknown element Z has a mass number of 31 and has 16 neutrons.

1 Name or give the true symbol for this element. *(Hint 1)*

2 Write its electronic structure. *(Hint 2)*

3 What charge ion would you expect it to form? *(Hint 3)*

4 If you spill a small amount of petrol, you can soon smell it all over the garage. Explain. *(Hint 4)*

5 Give the charge and electronic structure of the ions you would expect for: *(Hint 5)*
   a) Lithium
   b) Sulphur

A student wishes to find out what types of structure and bonding were present in three unknown substances, X, Y and Z

	X	Y	Z
Conducts electricity when melted?	No	No	Yes
Melting point	Very high	Low	High

6 Which substance has ionic bonding? Explain how you worked this out. *(Hint 6)*

7 Which substance has a simple molecular structure? Explain how you worked this out. *(Hint 7)*

Hints and answers follow

Atomic structure and bonding

Hints

1. Work out the number of protons.

   Find the atomic number on the periodic table.

2. Number of electrons = number of protons.

   Remember how many electrons go into each shell.

3. Will it lose or gain electrons?

   Remember – losing electrons gives a positive charge, gaining them gives a negative charge.

   Remember – it wants to get a full shell!

4. If you can smell it, some of the petrol particles must have reached your nose!

   How do liquids and gases spread out?

5. Find them in the periodic table.

   Remember – the group number tells you how many outer electrons you've got.

   How many electrons will each of them end up with when they've formed an ion?

6. What physical properties do ionic substances have?

7. What physical properties do simple molecular substances have?

Answers

1 Phosphorus (p) **2** 2,8,5 **3** P^{3-} **4** some of the petrol vaporises; the vapour diffuses throughout the garage **5** a) Li: electronic structure 2 b) S^{2-}: electronic structure 2,8,8 **6** Z – it's the only one that conducts electricity when melted (and it has a high melting point) **7** Y – it's the only one with a low melting point

Rates of reactions

Test your knowledge

1 Chemical reactions occur when particles _____ with enough _____ to form new substances.

2 The rate of all chemical reactions can be increased by increasing the _____ or _____ . For solids the rate of reaction can be increased by increasing the _____ _____ .

3 Calcium carbonate reacts with acid to form carbon dioxide gas. The rate of reaction can be found by measuring the _____ of _____ at intervals over a period of time.

4 The rate of reaction between calcium carbonate and acid can be compared for two different experiments by plotting a graph of _____ on the y-axis and _____ on the x-axis.

5 Calcium carbonate reacts faster with concentrated acid than dilute acid. If the same amount of calcium carbonate is reacted with dilute and concentrated acid, would the reaction with concentrated acid produce more, less or the same amount of carbon dioxide once the reaction has finished?

_____

6 A catalyst is a substance which _____ the rate of reaction, but which is not _____ _____ during the reaction.

7 Catalysts in living cells are called _____ .

Answers

1 collide / energy 2 temperature / concentration / surface area 3 volume or mass / gas 4 volume of gas / time 5 same 6 increases / used up 7 enzymes

If you got them all right, skip to page 70

Rates of reactions

Improve your knowledge

20 minutes

1. Chemical reactions occur when particles **collide** and form new substances. These collisions must occur with **enough energy** to be successful. The more often successful collisions occur, the faster the **rate of the reaction**.

2. The rate of a reaction can be increased by increasing several factors as shown below.

Temperature	Particles move more quickly **and** collide with more energy
Concentration*	More particles present so they are more likely to collide
Surface area of solids	Easier for other particles to come into contact with the solid

*Increasing the **pressure** of **gases** has the same effect

3. The rate of reaction can be found by measuring how much of one of the **products is formed** or how much one of **the reactants disappears** at intervals over a period of **time**.

4. The **results** of experiments can be **plotted on a graph**. For example the reaction between magnesium and dilute hydrochloric acid produces hydrogen gas. The graph shows the amount of hydrogen gas produced over a period of time for two experiments. In each experiment the same amounts of magnesium and acid were used. However, experiment 2 was carried out at a higher temperature.

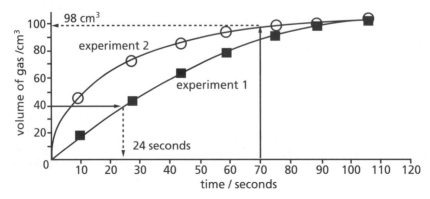

5 We can use the graph to find out information about the reactions as shown.

a) In experiment 1, we can see that 40 cm³ of hydrogen gas would be produced in 24 seconds. In experiment 2, after 70 seconds, 98 cm³ of hydrogen gas would be produced.

b) The line for experiment 2 rises **more steeply** than the line for experiment 1 at the start and this means that the reaction in experiment 2 has a **faster rate**.

c) However, the **same** total volume of hydrogen gas was produced in both experiments because the **same** amount of magnesium and acid were made to react together.

6 A **catalyst** is a substance which **speeds up a chemical reaction** but which is **not actually used up** in the reaction itself. Catalysts are widely used in large-scale industrial reactions because they speed up the reaction without having to use high temperatures or concentrations which would be more **expensive**.

7 In nature, living cells use catalysts called **enzymes**. For example, **yeast cells** and **bacteria cells** produce enzymes which are used in various industries as shown below:

Source of enzyme	Use	What the enzyme does
Yeast	Brewing	Converts sugar into alcohol. This is called **fermentation**
Yeast	Baking	Enzymes produce CO_2 gas which causes bread to rise
Bacteria	Dairy	Enzymes in bacteria cells convert milk to yoghurt

Enzymes will only work if they have the **right conditions**. If conditions are too hot or strongly acidic or alkaline, the enzyme stops working and the reaction slows down and stops.

Now learn how to use this knowledge

Rates of reactions
Use your knowledge

20 minutes

The graph below shows the results of two experiments to measure the rate of reaction between zinc granules and dilute hydrochloric acid. Both experiments were carried out using exactly the same mass of zinc, volume of acid and at the same temperature. However, the concentration of acid used was different in the two experiments.

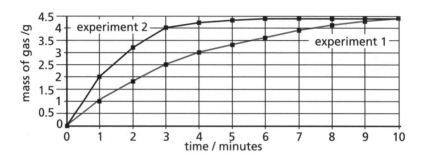

1 From the graph, find out how much hydrogen gas was produced in each experiment after 3½ minutes:

Experiment 1 _____   Experiment 2 _____

Hint 1

2 From the graph, find out how long it took in each experiment for 4 grams of hydrogen to be produced:

Experiment 1 _____   Experiment 2 _____

Hint 2

3 From your answers in 1 and 2, which reaction occurred at the faster rate?

Hint 3

4 Which experiment used the more concentrated acid? Explain your answer in terms of how reactions occur.

Hint 4

Many biological reactions are catalysed by enzymes. The graph shows how the rate of an enzyme-catalysed reaction changes as temperature varies.

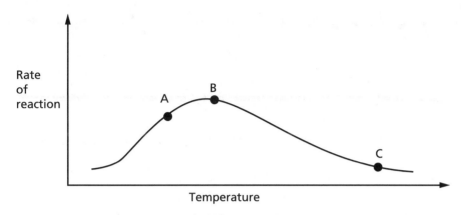

5 Between points A and B temperature increases. What happens to the rate of reaction? *(Hint 5)*

6 Explain this change in terms of the particles involved in the reaction. *(Hint 6)*

7 Between points B and C temperature increases. What happens to the rate of reaction? *(Hint 7)*

8 Why do you think this change occurs as the temperature increases? *(Hint 8)*

9 A chemical company, Alco, sets up a factory to produce a new plastic. Alco must produce enough plastic each week to be profitable. Alco's process uses an expensive catalyst. *(Hint 9)*

Why do you think Alco uses a catalyst to be profitable?

10 Why doesn't an expensive catalyst lead to high costs every week? *(Hint 10)*

Hints and answers follow

Rates of reactions

1. Draw a line from axis showing 'time' to the two curves and then draw a line to the axis showing 'amount of gas'.

2. As above but start from the 'amount of gas produced' axis.

3. In which experiment was hydrogen gas produced most quickly?

4. How does concentration affect the rate of reaction?

5. Read this from the graph.

6. As temperature increases, what happens to the speed of the particles?

7. Read this from the graph.

8. What happens to enzymes at high temperatures?

9. The plastic must be produced quickly enough. What does a catalyst do?

10. What happens to a catalyst during a chemical reaction?

Answers

1 experiment 1: 2.65 – 2.85 g / experiment 2: 4.0 – 4.2 g **2** 2.9 – 3.1 minutes / 7.1 – 7.3 minutes **3** experiment 2 **4** experiment 2 – more particles present therefore more collisions occur leading to a higher rate of reaction **5** increases **6** particles move more quickly and with more energy and so there are more successful collisions which speeds up the reaction **7** decreases **8** enzymes are damaged by high temperature so the reaction is not catalysed and slows down **9** a catalyst speeds up the reaction, producing more plastic per week **10** catalysts are not used up in a chemical reaction and so only need to be bought once

Rocks and plate tectonics

Test your knowledge

10 minutes

1 There are three types of rock: igneous, _____ and metamorphic.

2 Igneous rocks form when _____ cools and solidifies.

3 Sedimentary rocks can be recognised because they are built up in _____ . The sediments are compressed or _____ to form sedimentary rock.

4 Metamorphic rocks are formed from igneous and sedimentary rocks which are changed by _____ or _____ or both.

5 Only _____ rocks contain fossils. This is because fossils are destroyed by the heat and pressure when _____ and _____ rocks form.

6 The Earth's crust is split into huge sections called _____ , which can move. This is known as the theory of _____ _____ .

7 Volcanoes are often found at the _____ of plates which are moving. When plates collide one may sink below the other and _____ .

8 The breakdown of rock by air and water is called _____ .

Answers

1 sedimentary **2** magma **3** layers/strata **4** consolidated **4** heat / pressure **5** sedimentary / igneous / metamorphic **6** plates / plate tectonics **7** edges / melt **8** weathering

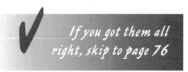
If you got them all right, skip to page 76

Rocks and plate tectonics

Improve your knowledge

1. There are three rock types: **igneous**, **sedimentary** and **metamorphic**, which are classified according to how they are formed.

2. Igneous rocks form when **magma** (molten rock) cools and solidifies.

3. Sedimentary rocks are formed from fragments of igneous or metamorphic rocks or from fragments of shells which build up in layers over millions of years. Eventually the great pressure turns the sediments into rock, i.e. consolidates them. The deepest sediments were buried first, i.e. they are the oldest.

4. Metamorphic rocks are formed from igneous or sedimentary rocks which are changed by heat or pressure or both.

5. Sedimentary rocks often contain fossils – but fossils are not found in igneous or metamorphic rocks – they are destroyed by the heat and pressure when the rocks are formed.

6. The Earth's crust is split into gigantic sections called **plates**. These float on the molten material below and can move apart or collide. This is the theory of **plate tectonics**.

7. When two plates collide, one of them may be pushed under and melt. The rock is therefore recycled back into magma. If one plate rides up over the edge of another plate or if neither plate sinks, the rocks are pushed upward and can form huge mountains, e.g. the Himalayas. Volcanoes and earthquakes are also common at the edge of tectonic plates. If two plates move apart, magma may be released violently in a volcanic eruption – it cools to form igneous rocks.

8 All rocks are slowly broken down by air and water (**weathering**). For example, a rock may be dissolved by acid rain or it may be slowly worn away by being rolled along in a fast-flowing stream. Weathering produces tiny particles of rock which can be blown or washed away, ending up as sediments.

The relationship between igneous, sedimentary and metamorphic rocks is shown by the **rock cycle**.

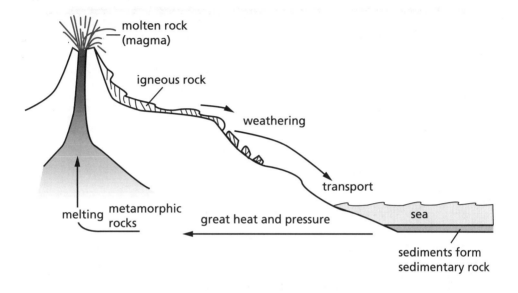

Now learn how to use this knowledge

Rocks and plate tectonics

Use your knowledge

The diagram shows the movement of three tectonic plates and the position of the South American and African continents.

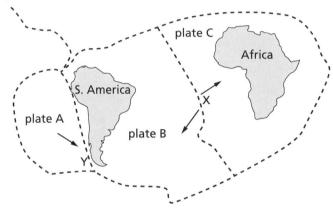

1 What would you expect to happen at X? (Hint 1)

2 What type of rocks would you expect to form at X? (Hint 2)

3 Plate A is colliding with and being forced below plate B. Why will metamorphic rock form below area Y? (Hint 3)

4 Why would you not expect to find fossils in the rocks below Y? (Hint 4)

5 How can an igneous rock on a mountain in Scotland end up as a sedimentary rock below the North Sea? (Hint 5)

Hints and answers follow

Rocks and plate tectonics

1 In what direction are the two plates moving?

What features do you often find when plates move like this?

2 Molten material will spurt up from below the sea. What kind of rocks form when this occurs?

3 What will happen to plate A when it is forced under plate B?

4 What conditions are needed for metamorphic rocks to form?

5 What happens to all rocks which are above the ground level?

How can rock particles be transported from one area to another?

How are sediments or particles turned into rock?

Answers

1 release of magma / volcanic activity **2** igneous **3** there are conditions of great heat and pressure **4** they will be destroyed by the heat and pressure **5** the rock will be weathered; the particles may be washed into rivers by rain; the rivers carry the particles to the sea where they build up as sediments

Metals and the reactivity series

Test your knowledge

10 minutes

1 Metals are _____ conductors of _____ and _____ . They have _____ bonding in their compounds. Their oxides and hydroxides are _____ .

2 Group 1 metals have _____ melting points. They react _____ with water to make a _____ and _____ . They get _____ reactive as you go down the group. Their compounds _____ in water.

3 Transition metals have _____ melting points. They are less _____ than group 1. Their compounds are _____ . Copper is used for wiring because it is a good _____ of _____ and can easily be _____ into wires. Iron is used as a building material because it is _____ .

4
a) Which is more reactive, aluminium or calcium?
b) Would you expect zinc to react with an acid?
c) Rubidium is more reactive than potassium. What method must be used to extract it from its compounds?

5 Will there be a reaction between zinc and magnesium chloride solution?

6 Iron is extracted from its _____ using the _____ _____ Here, _____ _____ , _____ and _____ are added at the top and _____ is blasted in at the bottom. The purpose of the _____ is to reduce the iron oxide to iron. The purpose of the _____ is to react with the impurities to form _____ which can be used on road surfaces. The iron formed is not very useful because it is _____ . Most of it is turned into _____ .

Answers

1 good / heat / electricity / ionic / basic **2** low / vigorously / hydroxide / hydrogen / more / dissolve **3** high / reactive / coloured / conductor / electricity / stretched / strong **4** a) calcium b) yes c) electrolysis **5** no **6** ore / blast furnace / iron ore / coke / limestone / air / coke / limestone / slag / brittle / steel

If you got them all right, skip to page 81

Metals and the reactivity series

Improve your knowledge

20 minutes

1 All metals conduct heat and electricity well. They are shiny. Metals react with non-metals such as oxygen and chlorine to form compounds with ionic bonding. All metal oxides and hydroxides are **basic**. The metal oxides and hydroxides that are soluble form **alkalis** when they dissolve in water. Most metals react with acids to give hydrogen gas.

2 Group 1 metals (lithium, sodium and potassium) are soft, and have a low melting point. They are very reactive, and must be stored in oil. The reactivity increases down the group. They look whitish, because their surface has reacted with the air. If you cut a piece, the cut surface will be shiny.

All group 1 metals react with water to make a solution of the metal hydroxide and hydrogen gas:

e.g.   sodium + water → sodium hydroxide + hydrogen

Compounds of group 1 metals all dissolve in water.

3 Transition metals (iron, copper, zinc) are hard and have high melting points. They are not very reactive – they hardly react with water at all.

Transition metal compounds are usually colourful – e.g. copper compounds are blue, iron compounds are green or brown.

Transition metals are useful – copper is used in wires because it conducts electricity well and it is easy to stretch it to make wires, and iron is used as a building material because it is strong.

4 Metals can be placed in order of reactivity. This gives the **reactivity series**. The reactivity of a metal tells you how it must be extracted from its compounds, as well as what it will react with.

Metal	Reaction with oxygen	Reaction with water	Reaction with acid	Extraction
Potassium	Tarnish in air Readily burn to form oxide	Violent reaction with cold water	Too violent to carry out	Electrolysis
Sodium				
Calcium		Reacts with cold water		
Magnesium		React with steam	React with dilute acids	
Aluminium				
Zinc	Burn slowly to form oxide			Blast furnace
Iron				
Copper		No reaction	Conc. acid needed	
Silver	No reaction		No reaction	Occurs as element

So zinc, for example, will burn slowly to make zinc oxide, will react with steam and dilute acids, and is extracted in the blast furnace.

5 A more reactive metal can displace a less reactive metal from its compounds in solution.

e.g.   zinc + copper sulphate solution → zinc sulphate solution + copper

BUT: if you added copper to zinc sulphate solution, nothing would happen, because copper is less reactive than zinc.

6 **Iron** is extracted from its ore **haematite** (iron oxide – Fe_2O_3) using the **blast furnace**.

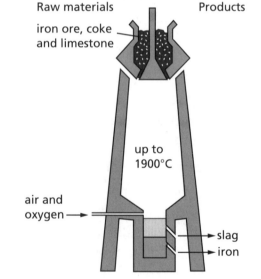

The **coke** produces **carbon monoxide**, which reacts with the haematite:

carbon monoxide + iron oxide
→ carbon dioxide + iron

The **limestone**, which is **basic**, reacts with **acidic impurities** to produce **slag**, which is used on road surfaces.

The iron produced is hard, but brittle, and is not very useful. Most of it will be turned into **steel**.

Now learn how to use this knowledge

Metals and the reactivity series

Use your knowledge

Caesium (Cs) is a group 1 metal. It is normally stored under oil to keep it fresh. A chemist found an old jar of caesium from which most of the oil had leaked. She noticed that the metal was covered in a white powdery substance.

1 Explain why caesium is stored under oil. *(Hint 1)*

2 What is the white powder and what is its formula? *(Hint 2)*

3 The chemist dissolved some of the white powder in water. What would happen if you put litmus paper into this solution? *(Hint 3)*

Noreen wishes to find out where the metal tin (Sn) comes in the reactivity series. She carries out several experiments:

4 She placed a piece of tin in cold water. Nothing happened. Where is the **highest** tin could be in the reactivity series? *(Hint 4)*

5 Noreen then put some more tin in a test-tube and added dilute hydrochloric acid. The metal dissolved, and gave off bubbles of gas.
   a) What is this gas, and how could you test for it?
   b) Where is the **lowest** tin could be in the reactivity series? *(Hint 5)*

6 Next, Noreen added a piece of tin metal to a solution of zinc sulphate. Nothing happened. What can she now conclude about tin's position in the reactivity series? *(Hint 6)*

7 Noreen needed to do one more experiment to find out exactly where tin is in the reactivity series. *(Hint 7)*
   a) Which other metal did she compare it with?
   b) Suggest an experiment she could have done to compare their reactivities.

Hints and answers follow

Metals and the reactivity series

Hints

1. You need to stop caesium reacting with anything while it is being stored.

   What might caesium react with if it was just in a jar on its own?

2. Since the oil drained away, the caesium has reacted with the air.

   What part of the air does it react with?

   What is the valency of metals in group 1?

   How can you use valencies to work out the formula?

3. What does litmus test for?

   What do you know about all metal oxides?

4. Look in the table about the reactivity series.

   It **must** be below all the metals that react with cold water.

5. What gas do you **always** get when a metal reacts with an acid?

   The test involves a lighted splint. What happens to it if the gas is present?

   It must be **above** any metal that doesn't react with dilute acids.

6. Remember a more reactive metal displaces a less reactive one.

   If there's no reaction, that tells you the metal in the compound is the more reactive of the two.

7. You should have worked out that tin must be either above or below one metal. Which one?

   There is only one reaction that allows you to directly compare two metals!

   You are looking for a displacement reaction – what chemicals would you have to use?

Answers

1 to stop it being in contact with the air and reacting 2 caesium oxide Cs_2O 3 it would turn blue (metal oxides are basic) 4 it must be below calcium 5 a) hydrogen – it will squeak/pop with a lighted splint b) it must be above copper 6 tin is less reactive than zinc, so it must be below zinc 7 a) we know it's below zinc and above copper – so need to find whether it's above or below iron b) add a piece of tin to a solution of a compound of iron (e.g. iron sulphate) (or alternatively, add a piece of iron to a solution of a compound of tin)

Non-metals

Test your knowledge

1. Give the colour and physical appearance of the following halogens.
   Chlorine _____ Bromine _____ Iodine _____ .

2. Halogens react with _____ to form compounds with ionic bonds. The halogen is present as an ion with _____ charge.

3. Hydrogen chloride contains _____ bonds. It dissolves in water to form a solution with an _____ pH.

4. Fluorine is more reactive than bromine. Therefore fluorine will _____ bromine from a solution of potassium bromide.

5. Chlorine can be tested for by using damp blue litmus paper. It first turns the paper _____ and then _____ it.

6. Chlorine is added to drinking water because it kills _____ .
   Fluoride is added to drinking water because it strengthens _____ .

7. Elements in group 0 all have the same physical state. What is this state? _____ . Because elements in group 0 already have 8 electrons in their outer shells they are _____ .

8. Neon is used in luminous _____ .

9. The test for hydrogen is that it _____ with a _____ _____ .

Answers

1 pale green gas / red-brown liquid / black solid 2 metals / –1 3 covalent / acidic 4 displace 5 red / bleaches 6 bacteria / teeth 7 gas / unreactive 8 signs or lamps 9 pops / lighted splint

If you got them all right, skip to page 86

83

Non-metals

Improve your knowledge

1. Elements in group 7 are called the **halogens**. They are non-metals with molecules containing two atoms. Some information about the halogens is shown in the table below.

Fluorine	Chlorine	Bromine	Iodine
F_2	Cl_2	Br_2	I_2
Pale yellow gas	Pale green gas	Red/brown liquid	Black solid*
Melting and boiling points decrease down the group			
Most reactive	Reactivity decreases down group		Least reactive

*Brown when in solution

2. Halogens react with metals to form **ionic** compounds containing the halogen as a **-1 halide** ion.

    e.g. $2K(s)$ + $Br_2(g)$ → $2KBr(s)$
    potassium + bromine → potassium bromide

3. Halogens react with other non-metals to form covalent compounds.

    e.g. $F_2(g)$ + $H_2(g)$ → $2HF(g)$
    fluorine + hydrogen → hydrogen fluoride

    The hydrogen halides (including hydrogen chloride, bromide and iodide) are very soluble in water and give **acidic** solutions which turn blue litmus paper red. The gases can be identified by testing with damp, blue litmus paper which turns red.

4. A more reactive halogen will **displace** a less reactive halogen from its compounds. For example, bromine will displace iodine from a solution of sodium iodide.

    $Br_2(l)$ + $2NaI(aq)$ → $2NaBr(aq)$ + $I_2(s)$

5 Chlorine can be identified in the laboratory by putting a piece of damp blue litmus paper in a tube containing the gas. The paper will first turn red and then white as it is **bleached**.

6 Some uses of halogens and their compounds are described below.

Element/Compound	Use	Reason
Fluorine	Toothpaste/drinking water	Strengthens teeth
Chlorine	Bleach Drinking water	Bleaching agent Kills bacteria
Iodine	Antiseptic	Kills bacteria
Silver halides, e.g. AgI	Photography	Change colour when exposed to light

7 Elements in group 0 (sometimes called group 8) are called the **noble gases**. Examples are helium, neon and argon. They are all very unreactive gases with very low boiling and melting points.

Because the noble gases all have 8 electrons in the outer shell they hardly ever react with other elements, and as elements the gases are made up of separate, unbonded atoms and not molecules. They are described as **monoatomic** (i.e. 'single atom').

8 Noble gases are used in street lamps and discharge tubes. A tube filled with gas has an electric current passed through it which causes a bright, luminous colour, such as in 'neon signs'.

9 Hydrogen is a non-metal that is not in a group. Its formula is H_2. It is explosive when mixed with air and a lighted splint will pop if hydrogen is present.

Now learn how to use this knowledge

Non-metals

Use your knowledge

Chlorine reacts with phosphorus to form phosphorus chloride.

1 What kind of bonding does phosphorus chloride contain? (Hint 1)

When phosphorus chloride reacts with water it gives off white misty fumes of hydrogen chloride which dissolve in water forming a solution. This solution reacts with potassium hydroxide solution, an alkali, to give potassium chloride and water.

2 Describe how you would test for hydrogen chloride gas and what you would see. (Hint 2)

3 Why does the solution react with potassium hydroxide solution? (Hint 3)

4 Write a word equation for this reaction including state symbols. (Hint 4)

Bromine reacts with calcium, Ca, to form a white solid W. W dissolves in water to produce a clear solution. When chlorine gas is bubbled through the solution a brown liquid, X, is produced and a new clear solution, Y.

5 What type of bonding is present in W? (Hint 5)

6 Identify W and give its formula. (Hint 6)

7 Explain what happens when the chlorine gas is bubbled through the solution of W, and therefore identify X and Y. (Hint 7)

8 Many years ago the water supply was not chemically treated and was not as safe as it is today. Diseases like cholera were much more common and many people suffered from tooth decay. Explain why the addition of chorine and fluorides has reduced the occurrence of these medical problems. (Hint 8)

The noble gases in group 0 are very unreactive. Because of this they exist as monoatomic gases and form very few compounds.

9 Explain what is meant by a monoatomic gas. *(Hint 9)*

10 How many electrons do noble gas atoms have in their outer shell? *(Hint 10)*

11 By thinking about their electronic structure explain why the noble gases are so unreactive. *(Hint 11)*

The gas hydrogen is very light and used to be used to fill 'hot-air balloons' and airships. However, it was not very safe as hydrogen burns in air releasing a lot of heat. These days helium is often used, even though it is not as light as hydrogen.

12 What do you think was the danger of using hydrogen? *(Hint 12)*

13 Why do you think helium is used instead of hydrogen even though it is heavier than hydrogen? *(Hint 13)*

Hints and answers follow

Non-metals

Hints

1 Phosphorus is a non-metal.

2 Learn this.

3 What type of solution do hydrogen halides form in water?

4 What are the reactants and products?

5 Ca is a metal.

6 What groups are the elements in – what are their valencies?

7 Chlorine is more reactive than bromine.

8 What causes the diseases and what do the halogens do?

9 'Mono' means *single*.

10 Look at the position of the group on the periodic table.

11 How many electrons do atoms want in their outer shells?

12 A **very large** amount of hydrogen was used and this could release a lot of energy.

13 What would happen if helium and air mixed?

Answers

1 covalent 2 place damp, blue litmus paper in the gas, it turns red 3 because hydrogen chloride solution is acidic and neutralises the alkali 4 aqueous hydrogen chloride + aqueous potassium hydroxide ← aqueous potassium chloride + liquid water (or solution instead of aqueous) 5 ionic 6 $CaBr_2$ 7 chlorine is more reactive than bromine and displaces it from its solution to form calcium chloride solution, Y, and bromine, the brown liquid X 8 chlorine kills bacteria in the water which cause diseases; fluorides strengthen teeth and help them resist bacteria in the water 9 a gas where the element exists as a single, unbonded atom 10 8 11 they already have a stable, full outer shell of electrons so they do not need to form any bonds with other atoms 12 it would be likely to explode or catch fire causing harm to people and damage to buildings, etc. 13 because it is unreactive it will not explode or react with air so it is much safer

Chemicals from oil

Test your knowledge

1 Crude oil is formed in the Earth's _____ by the long-term effects of _____ and _____ on the remains of organisms (marine deposits). Crude oil deposits are formed in _____ rock below _____ rock. Crude oil is a _____ of many different _____ . These are substances that contain _____ and _____ **only**. Often, where oil is formed _____ is also formed. As they are both _____ dense than water they rise up and collect below the _____ rock, and can be collected by _____ through this rock. The chemical name for natural gas is _____ . The chemical formula is _____ .

2 Crude oil can be separated by _____ _____ . This is done by _____ the oil and passing it through a _____ _____ . Hydrocarbons with a small number of carbon atoms in the molecule will boil at a _____ temperature. Larger molecules have a _____ boiling point. They are also _____ (i.e. don't flow very easily). Highly flammable hydrocarbons are ones with _____ molecules.

3 Burning hydrocarbons in excess oxygen (air) produces heat, _____ _____ , and _____ . The gaseous product turns lime water _____ . With a lack of oxygen, incomplete combustion occurs which produces _____ _____ and/or _____ as a product.

4 The burning of fuel often produces pollutants, e.g. _____ _____ which forms acid rain. Car exhausts often contain nitrogen oxides, which are produced by _____ and _____ being forced to react together by the high _____ in the engine.

5 Large hydrocarbons are broken down into smaller, more useful products by _____ . Some of these products are used as _____ , others are joined together to make very large molecules called _____ .

Answers

1 crust / heat / pressure / porous / non-porous/impermeable / mixture / hydrocarbons / carbon / hydrogen / natural gas/methane / less / non-porous / drilling / methane / CH₄ 2 fractional distillation / evaporating / fractionating column / low / high / viscous / small 3 carbon dioxide / water / milky (cloudy) / carbon monoxide / carbon 4 sulphur dioxide / nitrogen / oxygen / temperatures 5 cracking /fuels / polymers

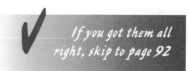

If you got them all right, skip to page 92

Chemicals from oil

Improve your knowledge — 20 minutes

1 Formation and extraction

Crude oil is formed in porous rock by the long-term effects of heat and pressure on marine deposits (dead animals/plants). Where oil is found, natural gas (methane) is also usually found. Oil and natural gas, being less dense than water, rise up through the porous rock and collect under non-porous rock.

2 Separation

The crude oil obtained by drilling through the non-porous rock is simply a mixture of lots of different **hydrocarbons**. Hydrocarbons are substances that contain hydrogen and carbon **only**. As with any mixture of liquids, the crude oil is separated by **fractional distillation**. Before separation the crude oil is heated until it evaporates.

The table below represents the properties of each of the different fractions obtained:

Name of fraction	Number of carbon atoms	Boiling range °C	Uses
Light petrol gases	1–4	<25	Portable fuels, e.g. camping gas
Petrol	4–12	25–60	Car fuel
Naphtha	7–14	60–180	Making other chemicals, e.g. fertilisers
Kerosine	9–16	180–220	Home heating (paraffin)
Diesel	15–25	220–250	Diesel car engines
Fuel oil	20–70	250–330	School central heating
Bitumen	>70	>350	Road surfaces

The bigger the hydrocarbon molecule, the higher its boiling point

3 Fuels

Many hydrocarbons are used as fuel, particularly methane (CH_4), otherwise known as natural gas, ethane (C_2H_6), propane (C_3H_8) and butane (C_4H_{10}). As well as providing energy (i.e. heat), combustion (burning) of fuels in a plentiful supply of air also gives carbon dioxide and water. If carbon dioxide is bubbled through limewater, the limewater turns 'milky'.

In a limited supply of oxygen, carbon monoxide and/or carbon is produced instead of carbon dioxide. Carbon monoxide is a colourless, odourless highly toxic gas.

4 Pollution

Combustion of fossil fuels (oil, gas, coal, etc.) is largely responsible for much of the pollution in the air. The table below shows some of these pollutants and their effects on the environment.

Pollutant	How it is formed	Effects on environment
Carbon dioxide	Combustion of fuels	Global warming
Sulphur dioxide	Burning of coal (sulphur is an impurity in coal)	Dissolves in rain water to form acid rain
Nitrogen oxides	In car engines, where high temperatures cause atmospheric nitrogen and oxygen to react	Form acid rain and smog

5 Cracking and polymerisation

Cracking is the process by which long chain hydrocarbons are broken down into smaller (more useful) hydrocarbons. Cracking is achieved by passing the hydrocarbon vapour over an aluminium oxide catalyst at 400°C.

Some of the small hydrocarbons formed during cracking are used as fuels, others undergo polymerisation. Polymerisation is the joining together of lots of small molecules to make one big one.

Examples and uses of polymers are given in the table below:

Polymer	Properties	Uses
Polythene	Light, flexible, strong	Carrier bags
Poly(propene)	Strong, rigid, light	Plastic chairs
Poly(styrene)	Light, insulator of heat	House insulation, packaging of electrical goods

Now learn how to use this knowledge

Chemicals from oil

Use your knowledge

20 minutes

The table below shows some of the fractions of oil and the boiling range of the constituent hydrocarbons.

Fraction	Boiling range °C
LPG	< 25
Petrol	
Naphtha	60–180
Diesel	220–250

1 Suggest a boiling range for petrol. *(Hint 1)*

2 Using information in the table what can you determine about the size of the molecules of petrol compared with the molecules of naphtha? *(Hint 2)*

3 Explain why petrol is considered to be more of a fire hazard than diesel. *(Hint 3)*

In an experiment, one of the hydrocarbons in the naphtha fraction was broken down to give one of the hydrocarbons present in the petrol fraction as well as another small molecule.

4 What is this process called and how is this carried out industrially? *(Hint 4)*

The other small molecule is joined together with lots of other molecules to form poly(ethene) {polythene}.

5 What name is given to this process? *(Hint 5)*

6 What properties does polythene possess that make it suitable to be used for plastic carrier bags? *(Hint 6)*

Fuel in a car engine is combusted (burnt) to release energy. The table below compares the composition of exhaust emissions from a new car and one that is 10 years old.

Old car	%	New car	%
Carbon dioxide	10	Carbon dioxide	53
Carbon monoxide	15	Carbon monoxide	1
Sulphur dioxide	17	Sulphur dioxide	6
Nitrogen oxides	58	Nitrogen oxides	40

7 A simple chemical test was used to detect the presence of carbon dioxide. What is the test and how does this show that carbon dioxide is present? *(Hint 7)*

8 One of the exhaust emissions is a colourless, odourless toxic gas. Which gas is this? *(Hint 8)*

9 One of the cars has been fitted with a catalytic converter. Which car do you think this is and what evidence in the table supports your answer? *(Hint 9)*

The table below gives the names and formulae of some hydrocarbons. Some of the information is missing.

Name of hydrocarbon	Formula
Methane	
	C_2H_6
Propane	
	C_4H_{10}

10 Complete the table. *(Hint 10)*

11 Which one of the above has the highest boiling point? *(Hint 11)*

12 Which one of the above is known as natural gas? *(Hint 12)*

13 What are the products of *complete* combustion of hydrocarbons? *(Hint 13)*

Hints and answers follow

Chemicals from oil

1. The petrol fraction is in between LPG and naphtha, so where will the boiling range be?

2. What is the connection between the size of a molecule and the boiling point?

3. Which fuel is more likely to turn into a vapour?

4. This is the process where long chain hydrocarbons are broken down into smaller (more useful) hydrocarbons.

5. Here, lots of small molecules are joining together to make a large molecule.

6. Properties are things like strength, density, how easily it can be made into shape.

7. Learn this!

8. This gas is formed when incomplete combustion occurs.

9. Which car gives off fewer pollutants?

10. All of these hydrocarbons are found in the LPG fraction.

11. What is the connection between molecular size and boiling point?

12. It is the smallest-sized hydrocarbon.

13. Learn these.

Answers

1 25–60°C (always put units in where appropriate) 2 petrol molecules must be smaller than naphtha molecules (because petrol has a lower boiling point) 3 petrol has smaller molecules than diesel hence petrol will turn into a vapour more easily (it is the vapours that are the fire hazard) 4 cracking, passing vapour at 400°C over an aluminium oxide catalyst 5 polymerisation 6 light, flexible, strong and easily made into shape 7 pass vapour through limewater and if it goes 'milky' then carbon dioxide is present 8 carbon monoxide 9 the newer car has the catalytic converter fitted because it emits lower levels of pollutants 10 reading left to right…CH_4, Ethane, C_3H_8, Butane 11 Butane (it has the largest-sized molecule of the four hydrocarbons) 12 methane 13 carbon dioxide, water and heat

Radioactivity
Test your knowledge

1. Alpha, beta, or gamma radiation is emitted when the _____ of an atom decays.

2. a) The most penetrating type of radiation is _____ radiation.
   b) The least penetrating type of radiation is _____ radiation.
   c) The type of radiation which will pass through a sheet of paper but is absorbed by 5mm of metal is _____ radiation.

3. The low level of radiation which is around us all the time is called _____ radiation.

4. The level of radiation can be measured using a _____ _____ .

5. What effects can radiation have on the cells in our body?

6. a) In what way can radiation be used in the treatment of cancer?
   b) What is the name of the radioactive isotope used to date fossil remains?

Answers

1 nucleus 2 a) gamma b) alpha c) beta 3 background 4 Geiger counter (or Geiger–Muller tube) 5 it ionises them, causing damage, cancer, or cell death 6 a) to kill cancerous cells (radiotherapy treatment) b) carbon–14

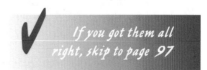

If you got them all right, skip to page 97

Radioactivity

Improve your knowledge

20 minutes

1. Some types of atoms have unstable **nuclei**. They are said to be radioactive and emit radiation.

2. **Alpha radiation (α)** can be stopped by a sheet of paper or a few centimetres of air. It consists of one or more helium nuclei.

   Beta radiation (β) is not stopped by air or paper, but most of it is stopped by a few millimetres of metal (e.g. foil). It consists of one or more electrons.

   Gamma radiation (γ) is the most penetrating of the three. Several centimetres of lead, or several *metres* of concrete will absorb most of it. Gamma radiation is part of the electromagnetic spectrum, with a very high frequency.

3. **Background radiation** is around us all the time. It is caused by radioactive substances in the Earth itself, in building materials, in the air, in food, even in our bodies, and from space. Some areas have higher levels of background radiation, due to the presence of rocks, such as granite.

4. Radiation can be detected using a **Geiger counter**. It is usually measured in counts per minute – the number of radioactive emissions entering the counter in one minute.

5. Radiation can be dangerous to humans. This is because when radiation collides with neutral molecules it can cause them to become charged or **ionised**. If the molecules in our body cells are ionised, they may be damaged, or even become cancerous. Larger doses of radiation kill cells.

6. Radioactive substances have a variety of uses:
   a) **Radiotherapy** uses radiation to kill cancer cells.
   b) **Radioactive tracers** can be introduced into the body, and their path through the body detected from outside. This enables certain medical problems to be diagnosed without an operation.

Now learn how to use your knowledge

Use your knowledge

An experiment to investigate a radioactive source is set up as shown below:

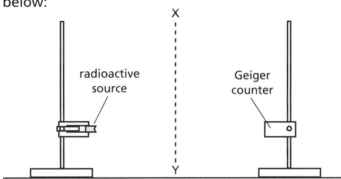

Different materials, such as paper and foil, may be put between the source and the counter, in the position marked X–Y.

The following readings were taken:

	Counts per minute
With no source present	6
With source, as shown in the diagram	98
With source, and a sheet of paper	36
Clamped between X and Y with source, and a 3mm thick sheet of aluminium clamped between X and Y	36

1 With no radioactive source present, the Geiger counter does not read zero. Why not? *(Hint 1)*

_____

_____

2 What is the radioactivity, in counts per minute, due to the source alone? *(Hint 2)*

_____

3 Some of the radiation is stopped by a single sheet of paper. What type of radiation is this likely to be? *(Hint 3)*

_____

4 No further radiation is cut out by the aluminium sheet. However, the count rate is still significantly above the original reading of 6 counts per minute. What other type of radiation does this suggest might be present? *(Hint 4)*

_____

5 Give 2 precautions that should be taken when carrying out this experiment.

1) _____

2) _____

6 Why are precautions needed when dealing with radioactive substances?

_____

_____

7 List 2 uses of radioactive materials.

1) _____

2) _____

Hints and answers follow

Radioactivity

Hints

1 There is always some radiation present around us.
What is it called? What causes it?

2 The 98 counts per minute is due to the source plus the background radiation. The background radiation count is the reading given before the source was put in place.

3 Only one type of radiation can be absorbed by a sheet of paper. Which one?

4 Both alpha and beta radiation would be cut out by this much aluminium. But some radiation is getting through the sheet. So what kind of radiation must be present?

Notice that adding the aluminium sheet didn't make any difference to the count rate. This means that the source can't be producing any beta radiation, as this would have mostly gone through the paper, but would then be stopped when the aluminium was used.

Answers

1 because of the background radiation from rocks, the air, our bodies, etc. 2 92 counts per minute 3 alpha radiation 4 gamma radiation 5 (e.g.) handle at arm's length, use tweezers 6 radiation can damage human cells and cause cancer 7 (e.g.) radiotherapy for cancer treatment, carbon–14 dating of fossils

Energy/energy transfer

Test your knowledge

1 A mug of hot coffee has _____ energy.

A radio gives out _____ energy.

2 The transfer of heat energy in a substance without that substance moving is called _____ . The transfer of energy in the form of a flow is called _____ .

3 Poor conductors are known as _____ .

4 An ordinary light bulb uses 100 J of electrical energy per second, and produces 5 J of light energy each second. A 'low energy' bulb uses 20 J of electrical energy per second, and produces 5 J of light energy each second. Which bulb is the most efficient?

5 A car moving at 60 mph will take _____ time to stop than a car moving at 30 mph, because the faster car has more _____ energy.

6 A boy lifts a pile of books off the floor onto a shelf 2 metres high. The force needed to lift them is 60 N.

a) Calculate the work done by the boy.

b) How much energy does the boy transfer to the books?

7 Two boys of the same weight ride identical bikes up the same hill. The journey takes Tim 4 minutes, but Jack gets there in 3½ minutes. Which boy produces the most power in his climb?

Answers

1 heat / sound **2** conduction / convection **3** insulators **4** the 'low energy' light bulb is more efficient **5** more / kinetic **6** a) 120 J b) 120 J **7** Jack

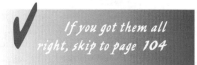

If you got them all right, skip to page 104

Energy/energy transfer

Improve your knowledge

1 Energy comes in many different forms. The main ones are:

heat energy (e.g. from the Sun)
light energy (e.g. from a light bulb)
sound energy (e.g. from a radio)
chemical energy (e.g. in a battery)
elastic or strain energy (e.g. in a stretched spring)
electrical energy (e.g. used by a TV set)
gravitational energy (due to height – e.g. bike at top of hill)
kinetic or movement energy (e.g. a car)
nuclear energy (e.g. used in a power station)

Energy is measured in joules (J).

2 If different parts of a substance are at different temperatures, heat energy will be transferred from the hotter part to the cooler part. Transfer of heat can occur by **conduction**, **convection** and **radiation**.

	Description	Example
Conduction	Transfer of heat energy in a substance without the substance moving	Heat transfer in metals
Convection	Transfer of heat energy by a flow	Heat transfer in liquids and gases
Radiation	Transfer of heat energy in rays	Objects emit radiation

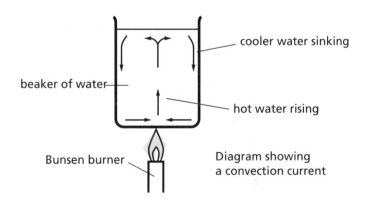

Diagram showing a convection current

3 Insulators reduce heat transfer because they are poor conductors. Often, insulating materials work by trapping air, a very poor conductor. This is used in cavity wall insulation (Wall B).

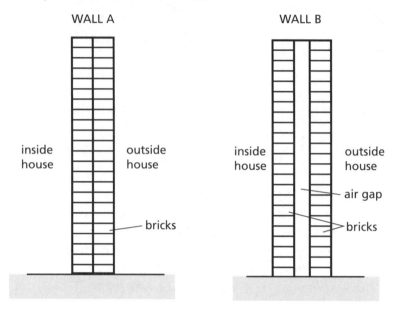

4 Energy can't be created or destroyed, but it can be changed from one form to another.

Energy transfers are not perfect, and some energy is wasted as **non-useful energy**, e.g. in a television set a little energy is turned into heat, and the back of the set feels warm. The less energy is wasted, the more efficient a device is.

The **efficiency** of a device is the fraction of the energy put in which is transferred to useful work.

$$\text{efficiency} = \frac{\text{useful energy out}}{\text{total energy in}}$$

you must be able to use this equation

5 Examples:

A moving car has kinetic energy. If the brakes are applied, the kinetic energy is turned into heat by the rubbing of the brake pads. The faster a car is going, the more kinetic energy it has, so the longer it takes for the car to stop.

An electric motor converts electrical energy into movement, e.g. in a hair dryer, fan heater, hoover, or washing machine. The wasted energy is usually some heat and sound energy.

6 Imagine we push a book a distance d along a table, and the force we need to push it with is F. The energy we've had to use (or the energy transferred from us to the book) is called the **work done**. It is measured in joules.

   work done = force (F) x distance moved (d)

   work done = energy transferred

   learn these

7 We could, of course, move the book quickly or slowly. Power is the rate of doing work – the quicker we do it, the greater the power.

$$\text{power} = \frac{\text{work done}}{\text{time taken}}$$

you must be able to use this equation

Now learn how to use your knowledge

Energy/energy transfer

Use your knowledge

1. Josh, a keen skier, is sitting in a ski lift which is moving at a constant speed up a mountain. What type of energy is he gaining as he moves higher up the slope?

Hint 1

2. Describe the energy changes that take place when he leaves the top of the mountain and starts to ski down. Try to include both useful and non-useful types of energy.

Hint 2

While eating lunch in the ski café, Josh looks at some information about the 2 ski lifts on that slope:

> Height of summit = 300 m
> (above base of lifts)
>
> Time to reach summit on blue lift = 3 minutes
>
> Time to reach summit on red lift = 3½ minutes

3. Josh knows that the force needed to lift him is 750 N (his weight). What is the work done in taking Josh from the base of the lifts to the top of the mountain?

Hint 3

4 power = $\dfrac{\text{work done}}{\text{time taken}}$

What is the power required to lift Josh up the mountain in 3 minutes?

_____

_____

5 The red and blue lifts are identical, except for their colour, and the times they take to reach the summit. Which lift is more powerful?

_____

6 efficiency = $\dfrac{\text{useful energy out}}{\text{total energy in}}$

Back in his chalet, Josh boils a kettle to make some coffee. The kettle uses 300 kJ of electrical energy, and the water gains 180 kJ of heat energy.

What is the efficiency of the kettle?

_____

_____

7 Suggest what may have happened to the rest of the energy (the missing 120 kJ).

_____

_____

_____

Hints and answers follow

Energy/energy transfer

Hints

1. As he's moving at constant speed, he isn't gaining kinetic energy.

   What type of energy does an object have due to its height?

2. What type of energy does he have when at the top?

   Which 2 types of energy does he have while skiing half way down?

   What type of energy does he have when he reaches the bottom (at high speed!)?

   What other types of energy may be produced, which may slow him down a bit?

3. The distance moved is just the height of the mountain, as the force is needed to lift him up, not to move him across.

   You must learn the equation for work done.

4. The time must be in seconds to give an answer in watts.

   3 minutes is 3 x 60 seconds.

5. Remember, the faster you do the work, the more powerful you are!

6. What else may be heated apart from the water?

Answers

1 gravitational energy 2 he loses gravitational energy, which is transferred to kinetic energy (useful) and a little heat and sound due to friction (non-useful) 3 225,000 J 4 1250 W 5 blue 6 0.6 (or 60%) 7 some energy is used to heat the kettle itself and the air around it, and a little sound is produced

Energy resources

Test your knowledge

1. Coal, oil and nuclear fuels are examples of energy resources which are limited and will one day run out. They are known as _____ energy resources.

2. Solar energy, the wind and the tides, however, will not run out, and are known as _____ energy resources.

3. a) Give one advantage and one disadvantage of coal as an energy resource.

   b) Give one advantage and one disadvantage of the wind as an energy resource.

4. In a coal-fired power station, coal is burnt to heat water. The steam produced turns the _____ , which in turn drive the generator, which produces _____ .

Answers

1 non-renewable 2 renewable 3 a) advantage: one of the following: currently easily available / fairly cheap and easy to convert to electricity / disadvantage: one of the following: supplies are running out / pollutes atmosphere, causing acid rain and increasing greenhouse effect. b) advantage: non-polluting / cheap / disadvantage: many turbines required for a small amount of energy / spoils large area of countryside / not reliable / only works if windy / only suitable in areas with reasonable winds 4 turbines / electricity

If you got them all right, skip to page 110

Energy resources

Improve your knowledge

1 **Fossil fuels** (coal, oil and gas) are the main energy resources used by humans. They are mostly burnt in power stations to produce electricity. Only a certain amount of these fuels exists, so they are known as **non-renewable** energy resources. Once used up they cannot be replaced as they take millions of years to form. Uranium, used as fuel in nuclear power stations is also a non-renewable energy resource.

2 **Renewable** energy resources are those which will not run out, or can be easily replenished. Examples are solar power, wind power, tidal power, hydroelectric power (from fast-flowing water), geothermal power (using the internal heat of the earth) and wood (trees grow quickly enough to replenish supplies).

3 There are a number of advantages and disadvantages to each of these energy resources.

 a) Non-renewable energy resources should be used sparingly as they will not last for ever.

 b) Burning fossil fuels in power stations causes pollution. Sulphur dioxide (a cause of acid rain) and carbon dioxide (which increases the greenhouse effect) are produced. There is also concern that large-scale projects such as 'wind farms' (large groups of wind turbines) spoil the landscape.

 c) Some energy resources are more reliable than others: the sun and wind will not always be available when needed!

 d) Nuclear fuels must be treated very carefully as they are radioactive. Public opinion is often against their use, because of the risk of an accident. Nuclear power stations are relatively cheap to run, but very expensive to 'decommission' at the end of their lives to make them safe. Radioactive waste products are produced by nuclear power stations, and these are difficult to dispose of satisfactorily.

e) The costs of resources vary. The sun, winds and tides are free, whereas fossil fuels must be mined. However, the equipment needed to harness the sun, winds and tides is expensive, as very large-scale projects are needed to generate as much electricity as one coal power station.

4 Electricity is the most convenient form of energy for us to use. It is generated in power stations.

In a traditional power station, fuel (coal/oil/gas) is burnt to heat water. The steam produced is used to turn **turbines**. The turbines drive the **generators** which produce electricity.

In nuclear power stations the heat from nuclear reactions heats the water, and in geothermal power stations water is pumped through hot rocks underground. In both cases the steam produced turns the turbines to drive the generators.

In other types of power station steam is not used. The turbines are turned directly by the wind, the flow of water downhill, or the water flowing out at low tide. The turbines then drive the generators in the same way.

Solar cells can be used to convert the sun's energy directly to electricity. They are often used on satellites.

Now learn how to use your knowledge

Use your knowledge

The diagram shows a pumped storage hydroelectric scheme. When demand for electricity is high, water is released from the top dam and flows rapidly down the pipe. As it flows through the power station its energy is used to generate electricity. The water is then stored in the lower reservoir. (*When demand for electricity is low, the water is pumped back up to the top reservoir, ready for use again.*)

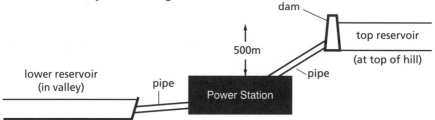

1 What type(s) of energy does the water have
a) when stored in the top reservoir? _____
b) while flowing down the pipe? _____

Hint 1
Hint 2

2 How is the flow of water used to produce electricity?
_____
_____

Hint 3

3 Other hydroelectric schemes do not pump water back up the hill, but use natural fast-flowing rivers. Give 2 advantages of this type of hydroelectric scheme, compared to coal-fired power stations.
_____
_____

Hints 4/5

Hints and answers follow

110

Energy resources

1. What kind of energy does something have if it is above the ground or up a hill?

2. What kind of energy does something have when it is moving?

3. Explain briefly how a power station, powered by flowing water, works.

4. What can you say about the cost?

5. How about pollution?

Answers

1 a) gravitational energy b) kinetic energy (and still some gravitational energy) 2 the flow of water turns the turbines directly, and the turbines drive the generators which produce electricity 3 the water is free, whereas coal is expensive to mine, and the hydroelectric scheme does not produce any polluting gases

Light and the electromagnetic spectrum

Test your knowledge

10 minutes

1 A ray of light hits a mirror at angle **a** and is reflected at angle **b**:

Is angle **a** bigger, smaller, or the same size as angle **b**?

2 A ray of light will change direction when it enters a glass block from the air, unless it hits the block at _____ .

3 Rays of light coming out of water into the air will be totally internally reflected if their angle with the normal is greater than the _____ _____ .

4 State one practical use of total internal reflection.

5 Describe what happens to white light when it goes through a triangular prism.

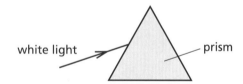

6 In the electromagnetic spectrum, which kind of wave has:
   a) the longest wavelength?
   b) the highest frequency?

7 Which type of electromagnetic radiation is used in grills, in radiant heaters, and to operate videos by remote control?

Answers

1 the same size **2** 90° (right angles) **3** critical angle **4** fibre optics or car reflectors or periscopes **5** the light is refracted (changes direction) and is separated into colours (dispersed) **6** a) radio waves b) gamma rays **7** infrared

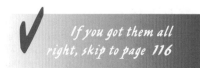

If you got them all right, skip to page 116

Light and the electromagnetic spectrum

Improve your knowledge

20 minutes

1 When light is **reflected** from a plane (flat) mirror, the angle at which it is reflected is the same size as the angle at which it hit the mirror:

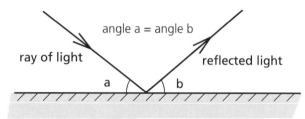

2 Light is **refracted**, i.e. it changes direction when it passes from one transparent substance to another. The speed of the light is different in different substances, and this causes the light to change direction (unless it hits the boundary at right angles i.e. at the **normal**).

Rays of light going from air into glass, water, or perspex bend *towards* the normal.

Rays of light going out of glass, perspex or water and into air bend *away from* the normal:

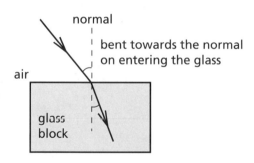
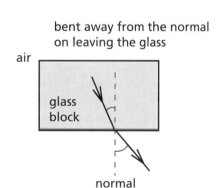

*the **normal** is always drawn at right angles to the surface, to help us measure the angles*

3 Rays of light coming out of glass, perspex or water into air will be **totally internally reflected** if their angle with the normal is greater than the **critical angle**:

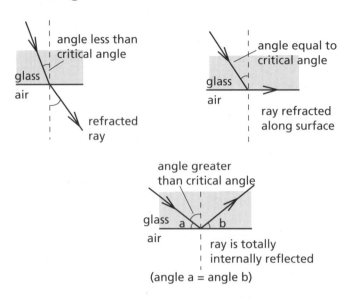

4 In optical fibres a ray of light is sent down a thin glass fibre. It cannot leave the fibre because if it hits the side it will be reflected by total internal reflection. If a fibre is passed into the body through the throat, a doctor can view the patient's stomach (for example), without needing to operate. This is called endoscopy.

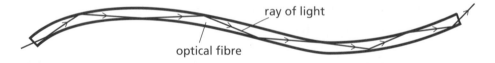

5 Ordinary light is called white light. It is made up of a mixture of all colours of light. These can be separated by a prism.

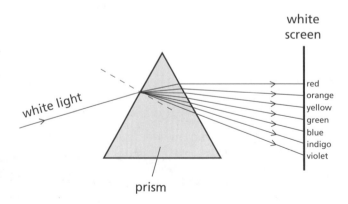

6 Light is one type of electromagnetic radiation. The electromagnetic spectrum consists of:

> gamma waves — shortest wavelength (high-frequency waves)
> X-rays
> ultraviolet light
> visible light
> infrared
> microwaves
> radio waves — longest wavelength (low-frequency waves)

increasing energy ↑   *increasing wavelength* ↓

7 **Gamma rays** are used in medicine to sterilise surgical instruments and to kill cancer cells.

X-rays pass through flesh but are absorbed by bone, so can be used to 'photograph' fractures.

Ultraviolet light is used in sunbeds to produce a tan.

Infrared light causes heating when absorbed by an object. It is used in grills, toasters and TV remote controls.

Microwaves are easily absorbed by water molecules in food, causing it to be heated and cooked.

Radio waves are used to transmit radio and TV signals.

Now learn how to use your knowledge

Light and the electromagnetic spectrum

Use your knowledge

20 minutes

A ray of light enters a perspex block at an angle as shown below:

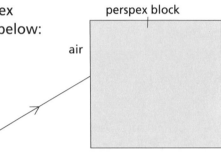

1 On the diagram, sketch the path of the ray inside the block. (Hint 1)

2 What is the name given for what happens as the ray enters the block?

_____

A ray of light passes through a glass prism as shown:

The critical angle for glass is 42°.

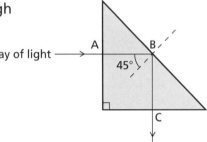

3 Why doesn't the ray of light change direction when it enters the prism? (Hint 2)

_____

4 State what happens to the ray at B. (Hint 3)

_____

Hints and answers follow

Light and the electromagnetic spectrum

1. First draw in the normal as a dotted line.

   Remember that rays *entering* perspex or glass are bent *towards* the normal.

2. At what angle does the ray hit the block at A?

3. What is the angle between the ray and the normal at B?

   The critical angle for glass is given in question 2.

Answers

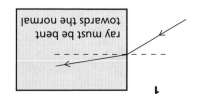

1 see diagram 2 refraction 3 it hits the prism at right angles (90°) 4 it is totally internally reflected

Waves and sound

Test your knowledge

1 Give one example of a transverse wave.

2 Give one example of a longitudinal wave.

3 The distance between 2 adjacent peaks on a water wave is called the _____ of the wave.

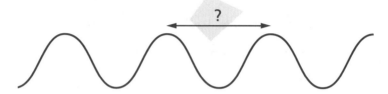

4 When water waves pass suddenly into shallower water they are slowed down. As a result of this they change direction. This is an example of _____ .

5 A _____ wave travels through the air as a vibration of the air molecules.

6 a) If the amplitude of a sound wave increases, it will sound _____ .
b) If the frequency of a sound wave increases, it will sound _____ .

7 What causes an echo?

8 Very high-frequency sound waves, used in the prenatal scanning of babies, are known as _____ .

Answers

1 light waves / water waves / waves in a rope 2 sound waves / some waves in springs 3 wavelength 4 refraction 5 sound 6 a) louder b) higher (in pitch) 7 the reflection of a sound wave 8 ultrasound

If you got them all right, skip to page 121

Waves and sound

Improve your knowledge

1 Waves made by shaking a rope, ripples in water, and light waves, are called **transverse** waves. They look a bit like this:

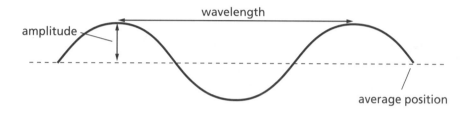

2 Some waves in springs, and sound waves, are a different kind of wave, called a **longitudinal** wave. They look a bit like this:

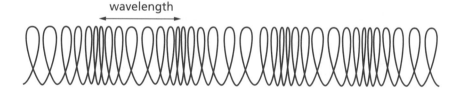

3 The **amplitude** of a wave is the maximum displacement from the average position.

The **wavelength** of a wave is the distance between two peaks of the wave.

The **frequency** of a wave is the number of cycles (up and down) it completes every second. It is measured in Hertz (Hz). The speed of a wave is found by multiplying together the wavelength and frequency. It is measured in metres per second (m/s).

4 All waves:
- can be reflected
- can be refracted
- transfer energy without transferring matter.

5 A sound is produced by a vibrating object – e.g. a violin string, or the voice box in our throat. It travels through the air as a wave, vibrating the air molecules.

6 The bigger the **amplitude** of a sound wave, the louder the sound.

The higher the **frequency** of the sound wave, the higher the pitch of the note.

7 A sound wave can be reflected from a hard, flat surface to produce an echo.

8 **Ultrasound** is a sound wave with a very high frequency, too high for us to hear. It is reflected by body tissues and can be used to scan a pregnant woman to check on her unborn baby.

✓ Now learn how to use your knowledge

Waves and sound

Jake bursts a balloon a little distance away from a brick wall. The noise it makes travels through the air as a sound wave. Jake hears an echo.

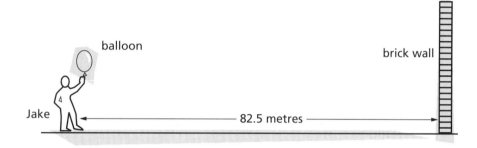

1 What happens at the brick wall that causes Jake to hear an echo?

_____

_____

2 There is a delay of 0.5 seconds between the balloon bursting and the echo reaching Jake. How far has the sound wave travelled in this time?

(Hint 1)

_____

3 Use the information given to calculate the speed of sound in air.

$$\text{speed} = \frac{\text{distance travelled}}{\text{time taken}}$$

_____

_____

4 Echoes can also occur with ultrasound. Give one use of ultrasonic echoes.

_____

_____

Some water waves are moving across a tank of water. The tank, viewed from the side, is shown below. The waves are moving to the right.

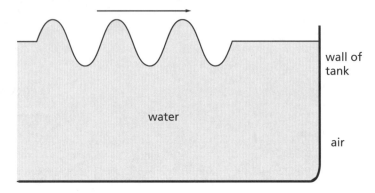

5 Mark on the diagram

    a) the wavelength of the waves.
    b) the amplitude of the waves.

(Hint 2)

6 What will happen to the waves when they hit the wall of the tank?

(Hint 3)

_____

_____

7 Is this wave more similar in nature to a light wave or a sound wave? Give a reason for your answer.

(Hint 4)

_____

_____

Hints and answers follow

Waves and sound

1. Remember that the sound wave has travelled to the wall, *and back*.

2. Remember that the amplitude is the greatest distance from the *average* position.

3. Remember the properties of waves.

   The situation is roughly the same as a sound wave hitting a wall or cliff, or a light wave hitting a mirror.

4. Is the water wave transverse or longitudinal?

Answers

1 the sound wave is reflected 2 165 m 3 330 m/s 4 finding the depth of water (or scanning unborn babies) 5 see diagram 6 they will be reflected back to the left 7 a light wave because it is transverse

123

Electricity

Test your knowledge

1 Draw the symbol for a resistor.

2 If the voltage across a resistor is 4V, and the current through the resistor is 0.4A, what is the resistance?

3 In a series circuit, the current is _____ _____ all around the circuit. In a parallel circuit, the _____ is the same across each of the branches.

4 To measure the voltage across a lamp, a _____ must be placed in _____ with the lamp.

5 In a 3-pin plug, the _____ wire is connected to the fuse. The _____ wire should normally have no current flowing through it.

6 You have the following fuses: 1A, 3A, 5A, 13A.

Which should be used in an appliance where the current is 6A?

7 What is the current flowing through a 100W light bulb when it is used with a 240V supply?

8 The current used in our homes is called _____ current, because it changes direction many times a minute. Batteries, however, use _____ current.

Answers

1 see diagram **2** 10Ω **3** the same / voltage **4** voltmeter / parallel **5** live / earth **6** 13A **7** 0.42A **8** alternating / direct

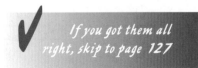

If you got them all right, skip to page 127

Electricity

Improve your knowledge

1 To understand an electric circuit, you need to know what the different symbols mean

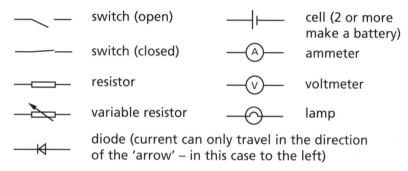

2 **Resistance** makes it harder for current to flow in a circuit. The units of resistance are **ohms** (Ω). When a current flows through a resistor, **heat** is produced. This is how electric heaters work.

Voltage, current and resistance are connected by the formula

Voltage = Current x Resistance   *learn this*

3 Electrical circuits can be connected in **series** or in **parallel**. In a series circuit, the current is the same all around the circuit.

2 lamps in series

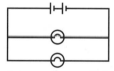
2 lamps in parallel

In a parallel circuit, the current splits when the circuit branches (If we add up the current in all the branches, it will be the same as the current leaving the battery.)

In a parallel circuit, the **voltage** is the same across each of the branches.

4 An ammeter must be placed **in series** to measure the current.

To measure the voltage, a voltmeter must be placed **in parallel**.

5 A standard electric plug

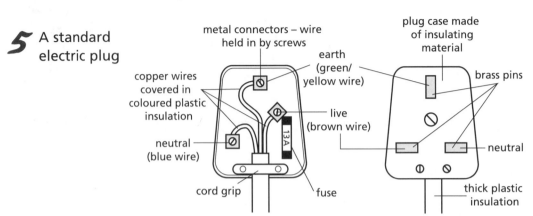

The current enters through the **live wire** and leaves the appliance through the **neutral wire**. Normally, no current flows through the **earth wire**, which is provided for safety, in case there is a fault and the metal case of an appliance becomes 'live'. The earth wire is connected to the metal case, and safely carries current away from it, which stops anyone getting an electric shock.

6 A fuse is a thin piece of wire in a plug which will melt if too much current flows into it – this stops the current. A 5A fuse, for example, would melt if a current of 5A or more passed through it.

7 The **power** of an appliance is measured in **watts (W)**.

Power = voltage x current   (learn this)

8 Energy used = power x time

If power is measured in watts and time in seconds, energy will be in joules.

If power is measured in kilowatts and time in hours, energy will be in kilowatt hours (units of electricity).

The electric current supplied by a battery is **direct current (d.c.)**, so current always flows in the same direction.

Mains electricity is **alternating current (a.c.)**. In the UK this changes direction constantly; it changes direction 50 times a second.

✓ *Now learn how to use your knowledge*

Electricity

Use your knowledge

Mary has a stainless steel electric kettle.

1 The power of the kettle is 2400W. Mary uses it with mains electricity, which is supplied at 240V. What current does her kettle use?

The plug is wired as shown here.

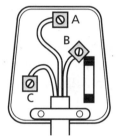

2 How much current flows through the wire attached to A?

3 Mary uses her kettle for 3 minutes a day. How much energy does it use each day, in joules?

Hint 3

4 In 3 months, Mary will have used her kettle for about 4½ hours. How much energy has been used in that time? (Give your answer in **kilowatt hours**.)

Hint 4

5 Electricity is 6p per unit. How much has it cost Mary to use her kettle over that 3-month period?

6 Mary's mother buys her a new, smart, plastic kettle. When Mary starts to wire the plug, she notices that there are only two wires, not 3.
  a) Which wire is missing?
  b) Why is this wire not needed for her new kettle?

Hint 7

Hint 8

7 Mary's new kettle uses a current of 11A. What is its resistance?

Hints and answers follow

Electricity

Hints

1. Write down the formula for power.

2. What is the name of this wire?

3. Write down the formula for energy. Remember to convert the time to seconds.

4. Use the same formula as for question 3. Remember the time must be in hours and the power in kilowatts. To convert watts into kilowatts, divide by 1000.

5. 6p per unit means 6p per kilowatt-hour.

6. Multiply 6p by the number of kilowatt-hours.

7. Which wires HAVE to be there for there to be a circuit?

8. What is the casing of the new kettle made of?

9. Write down the formula connecting resistance, voltage and current.

   Remember the voltage will be the same for the new kettle and the old kettle.

Answers

1 10A **2** none, because it is the earth wire **3** 2400 × 180 = 432,000J **4** Power in kW = 2400/1000 = 2.4, so energy in kWh = 2.4 × 4.5 = 10.8kWh **5** 10.8 units used, so cost = 10.8 × 6 = 64.8p **6** a) the earth wire b) because the case of the kettle isn't metal **7** resistance = voltage/current = 240/11 = 21.82W

Electromagnetism

Test your knowledge

1 Will the following pairs of magnets be attracted, repelled, or unaffected by each other?

a) [N S] [S N] _____

b) [N S] [N S] _____

c) [S N] [N S] _____

2 a) A coil of wire will behave as a magnet if a _____ is flowing through it.

b) A _____ _____ bar through the centre of the coil will increase the strength of the magnet.

3 What happens to an electromagnet when the current through it is turned off?

4 What happens to a wire with an electric current going through it, if it is put in a magnetic field?

5 What happens to a metal wire moving in a magnetic field?

6 a) What device is used to increase the voltage of electricity leaving a power station?

b) Why is it better to send electricity round the country at a high voltage?

Answers

1 a) repelled b) attracted c) repelled **2** a) current b) soft iron **3** it no longer behaves as a magnet **4** it has a force on it **5** a voltage is induced across it **6** a) transformer b) the current is lower, so less energy is lost as heat in the wires

If you got them all right, skip to page 133

129

Electromagnetism

Improve your knowledge

1 **Bar magnets** have a North pole and a South pole. Like poles repel. Opposite poles attract. Magnets have magnetic fields around them.

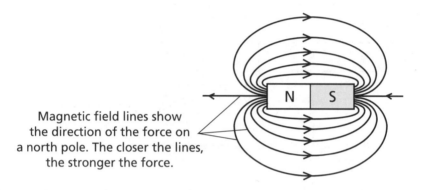

Magnetic field lines show the direction of the force on a north pole. The closer the lines, the stronger the force.

2 An **electromagnet** is just a coil of wire with an electric current going through it. It behaves just like a bar magnet when the current is switched on.

The strength of the electromagnet can be increased by:

a) increasing the current

b) using a coil with more turns of wire

c) putting a soft iron core through the centre of the coil.

Electromagnets are often useful because they can be turned on and off.

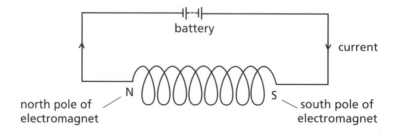

3 In an electric bell:

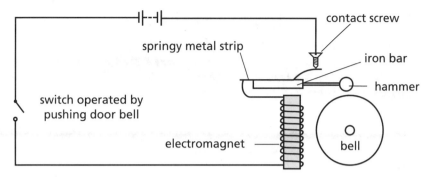

When the switch is closed, a current flows through the electromagnet, down the metal strip, through the contact screw, and back to the battery. The electromagnet attracts the iron bar. As the bar moves, the hammer hits the bell.

However, with the bar stuck to the magnet, the contact screw is no longer touching the metal strip, so there is a gap in the circuit and the current stops. The electromagnet turns off, so the bar springs back to its usual position.

With the iron bar back in its original position touching the contact screw, current flows again, and the process is repeated. So the hammer hits the bell over and over again as long as the door bell switch is being pressed.

4 If a wire with current flowing through it is put in a magnetic field, it will experience a force (unless it is parallel to the magnetic field lines). In an electric motor (*see below*), the left side of the coil (A) has a downward force on it (into the paper). The right side (B) has an upward force, as the current is flowing in the opposite direction. These forces make the coil rotate.

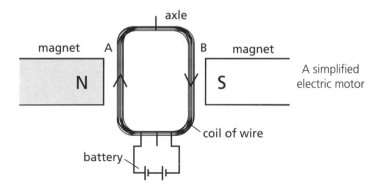

A simplified electric motor

131

5 A conductor (e.g. a metal wire) will have a voltage **induced** across it if:

1) the magnetic field around it is changing

2) it is moving through a magnetic field.

If the conductor is part of a complete circuit, then the induced voltage causes a current to flow in the circuit.

This principle is used to generate electricity in power stations, when a magnet rotates inside a coil of wire (or a coil of wire is rotated in a magnetic field).

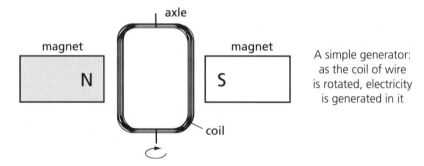

A simple generator: as the coil of wire is rotated, electricity is generated in it

6 **Transformers** are used to change the size of an a.c. voltage. At the power station they increase the voltage of the electricity before it is sent round the country on power lines. The voltage is reduced again by local transformers to make it safe for use in houses.

High voltages are used in the power lines because then only a low current is needed. This causes much less heating in the cables, so less energy is lost as heat.

Transformers only work with a.c. voltages, which is why a.c. is used for mains electricity.

Now learn how to use your knowledge

Electromagnetism

Use your knowledge

The diagram shows an electromagnet:

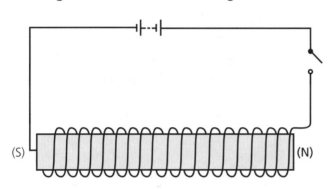

small compass

When the electromagnet is switched on it has North and South poles, marked (N) and (S). The compass pointer is a small permanent magnet, with poles marked N and S.

1 Describe what happens to the compass when the electromagnet is turned on. Hint 1

_____

_____

2 On the diagram, draw the shape of the magnetic field around the electromagnet. Hint 2

Electromagnetism

Hints

1. The compass is close to the North pole of the electromagnet, and the pointer is free to turn around. Which end of the compass pointer is attracted to the North pole of the electromagnet?

2. Remember a coil of wire behaves as a bar magnet when it has a current going through it. So, the magnetic field will be the same shape as for a bar magnet.

Answers

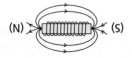

1 the South pole of the pointer swings round to point towards the North pole of the electromagnet 2 see diagram

Forces and motion

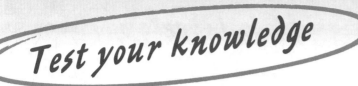
Test your knowledge

10 minutes

1 a) Calculate the average velocity of a car which travels 90 miles North in 2 hours. (speed) = d/t in direction

b) Calculate the acceleration of a car which increases its velocity from 10m/s to 30m/s in 10 seconds. acc^n = velocity/time

2 a) Which is moving faster, marble A or marble B?

b) Is the car speeding up, slowing down or moving at a constant speed?

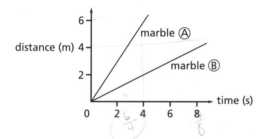
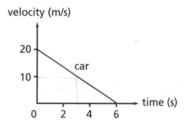

3 These three trolleys are each pulled by a force of 3N. Which one will have the greatest acceleration?

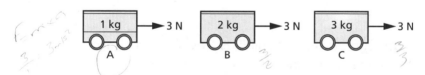

F = m×a
3 = 3m/s²

4 If balanced forces are acting on an object, its speed will be ___constant___.
A marble rolling across the floor stops because of ___friction___.

5 A man is leaning on a wall. He puts a force of 10N on the wall. What force does the wall put on the man?
10N → 0
if < or he would fall over
if > or it would fall over

Answers

1 a) 45 mph North b) 2m/s²   2 a) A b) slowing down   3 A
4 constant / friction   5 10N

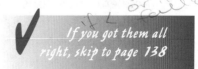
If you got them all right, skip to page 138

135

Forces and motion

Improve your knowledge

1 $$\text{speed} = \frac{\text{distance travelled}}{\text{time taken}}$$ *learn this*

Velocity is speed in a certain direction – e.g. 20m/s South.

If your velocity changes, you **accelerate**

$$\text{acceleration} = \frac{\text{velocity change}}{\text{time taken}}$$ *learn this*

Acceleration is measured in m/s^2.

2 A **distance–time graph** is a 'plot' of the object's motion.

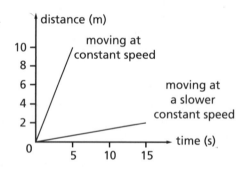

If it is a **straight line,** the object is travelling at constant speed.
The steeper the slope, the greater the speed.
If the line is horizontal, the object isn't moving.

3 A **velocity–time graph** shows how the velocity varies.

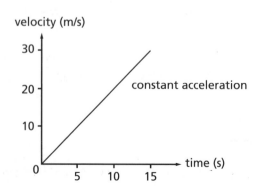

If it is a straight line, then the acceleration is constant.
If the line is horizontal, the velocity is constant.

Balanced forces are forces which cancel out.

e.g. 3N ←[book]→ 3N   no overall force acts

Friction is the force which acts to slow you down! It happens when two solid surfaces (like a tyre and the road) move across each other, or when something moves through air or water (air or water resistance).

The direction of the friction is always opposite to the direction the object is moving in.

 If **unbalanced forces** (forces which do not cancel out) act on an object, then it will **accelerate** (speed up or slow down).

e.g. friction 1100N   forward force due to engine 900N   The car will slow down because the resultant force is 200N to the left.

The bigger the force, the greater the acceleration.

But the bigger the mass of the object, the smaller the acceleration a particular force will produce.

When one object (e.g. a mug) puts a force on another object (e.g. a table), the second object (table) always puts an equal force, in the opposite direction, on the first object (mug).

e.g.   Force of mug on table = 1N (its weight)

So force of table on mug must be 1N too (this is called the **reaction force**).

1N ↑ ↓1N

Now learn how to use your knowledge

Forces and motion

Use your knowledge

The graphs show the movement of two different people over a 20-second time period:

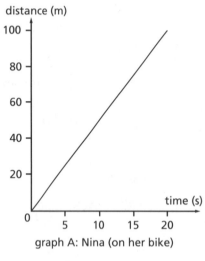

graph A: Nina (on her bike)

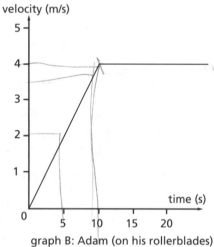

graph B: Adam (on his rollerblades)

1 Calculate Nina's speed.

Hint 1

2 Sketch a velocity–time graph for Nina's motion.

Hint 2

3 Now look at graph B. Describe Adam's motion:

a) in the first 10 seconds

b) between 10 and 20 seconds.

Hint 3

4 Find Adam's acceleration between 5 and 10 seconds.

Hint 4

A child's plastic football is dropped from the top of a tower block on a day when there is no wind. As it passes the top floor windows, the forces on it are as shown below, and its speed is 5m/s.

5 What is the cause of the downward force on the ball? *(Hint 5)*

6 What is the cause of the upward force on the ball? *(Hint 6)*

7 Describe the ball's motion at this point. *(Hint 7)*

8 By the time it has reached the second floor, the speed of the ball has stopped increasing and has become constant. Explain why this has happened. *(Hint 8)*

Hints and answers follow

Forces and motion

Hints

1. Speed is the gradient of a distance–time graph (or you could use speed = distance/time).

2. What is her velocity? Is it increasing, decreasing or constant? How do you show this on a graph?

3. What does 'describe' want you to do? What does a straight line on a velocity–time graph mean? If the line is horizontal, this is a special case.

4. What is the formula for acceleration?

5. Why does anything fall down?

6. What is the ball moving **through**? This force is trying to slow the ball down.

7. The forces aren't balanced. What happens when an object is acted upon by unbalanced forces?

8. What do you think will happen to the air resistance as the ball gets faster? How will this affect the ball's acceleration?

Answers

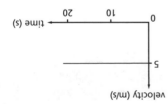

1 5m/s 2 see graph 3 a) constant acceleration b) constant velocity 4 0.4m/s² 5 weight (or gravity) 6 air resistance (or friction) 7 the ball is accelerating downwards 8 as the speed of the ball increased, the air resistance increased until it equalled the ball's weight; when this happened, the forces on the ball were balanced

The Earth and beyond

Test your knowledge

1. The Moon is constantly orbiting the _____.
   The Earth is constantly orbiting the _____.

2. The Sun, Earth, and other planets are together known as the _____ _____.

3. The Sun is our nearest _____. We can see the Moon because it _____ light from the Sun.

4. Our Sun is part of a group of millions of stars called the _____ _____.

5. The planets remain in orbit around the Sun because of the force of _____. This force depends on the mass of the planet, and also on its _____ from the Sun.

6. Man-made objects which orbit the Earth to send television signals around the world or to help with weather observations are called _____.

Answers

1 Earth / sun 2 solar system 3 star / reflects 4 Milky Way galaxy 5 gravity / distance 6 satellites

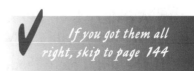

The Earth and beyond

Improve your knowledge

1 The Earth is constantly **orbiting** the Sun, following a path shaped like an ellipse (a slightly squashed circle). It takes a year for the Earth to go round the Sun once.

At the same time, the Moon is orbiting the Earth. It takes about 4 weeks to go round once.

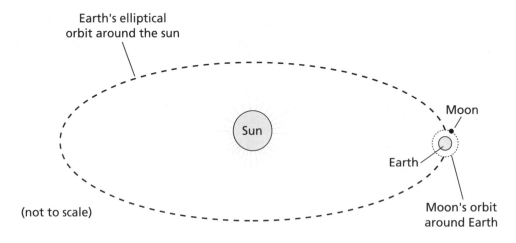

2 The **solar system** consists of the Sun, and the Earth and other planets (etc.) which orbit it. Some of the other planets also have Moons. In general, the closer a planet is to the Sun, the hotter it will be.

(See diagram on next page.)

3 The Sun is our nearest star.

A **star** gives out light and heat. **Planets** and **Moons** do not give out light, but we can see them if they reflect light from the Sun.

4 A **galaxy** is a vast group of stars. Our Sun is just one of millions of stars which make up the Milky Way galaxy.

The **universe** is made up of at least a billion galaxies.

5 The planets are held together in the solar system, and the stars are held together in the galaxy, by the force of **gravity**. This force acts between all objects with a mass – but it is stronger if the objects have a large mass, or are close together. Without gravity the planets would not orbit the Sun: they would just keep moving in a straight line through the galaxy.

6 Artificial **satellites** are man-made objects which can be made to orbit the Earth. They are used to monitor the weather and relay TV signals around the world.

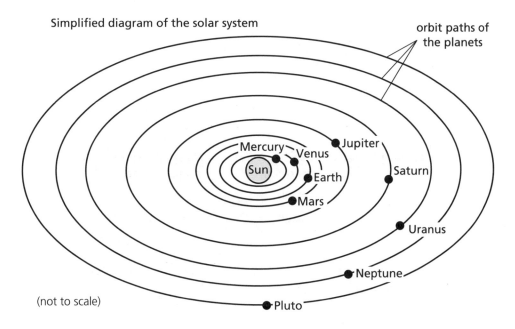

Simplified diagram of the solar system

orbit paths of the planets

(not to scale)

Now learn how to use your knowledge

143

The Earth and beyond

Use the following list to answer the questions below:

Earth, Moon, Milky Way galaxy, Sun, solar system.

1 Which of the above is:
   a) the largest? _____
   b) the smallest? _____
   c) a star? _____

(Hint 1)

Planet name	Distance from Sun (millions of km)	Mass of planet (compared to Earth)
Jupiter	780	320
Neptune	4500	18

2 Which of the two planets in the table would you expect to have the coldest surface temperature? Why?

_____
_____

(Hint 2)

3 Both planets are held in their orbits around the Sun by the force of gravity. Give 2 reasons why the force holding Jupiter in orbit is greater than that holding Neptune in orbit.

_____
_____

(Hint 3)

Hints and answers follow

The Earth and beyond

1 Re-read *Improve your knowledge* if you are unsure. You must know what each of these is.

2 Which piece of information in the table will affect how warm the planet is?

Planets get their heat (mostly) from the Sun, so the closer they are to the Sun the warmer they will be.

3 The force of gravity between two objects is larger the larger their masses, and the closer together they are.

The closer together the Sun and the planet are, the greater the force of gravity between them.

The greater the mass of the planet, the stronger the force of gravity between it and the Sun.

Answers

1 a) Milky Way galaxy b) Moon c) Sun 2 Neptune, because it is further from the Sun 3 Jupiter has a greater mass, and is closer to the Sun

145

Mock exam

1 The diagram below shows a simple food chain.

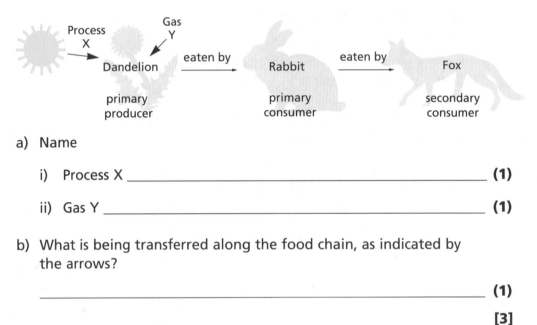

   a) Name

      i) Process X _____ (1)

      ii) Gas Y _____ (1)

   b) What is being transferred along the food chain, as indicated by the arrows?

      _____ (1)

                                                                    [3]

2 The table shows the composition of one slice of wholemeal bread.

Energy	392kj
Protein	4.0g
Carbohydrate	18.3g
Fat	1.1g

   a) One slice of bread contains 392kj. What does kj stand for?

      _____ (1)

b) Where in the body does protein digestion begin?

_____ (1)

In an investigation, a student weighed himself on Monday morning and then, for the next two days, weighed all of the food he consumed. He then weighed himself again on Wednesday morning. The student's recordings are shown below.

Weight (Monday)	28.5kg
Amount of food consumed	2.7kg
Weight (Wednesday)	28.5kg

c) Suggest why the student's weight did not increase significantly.

_____
_____
_____ (2)

[4]

3 a) Name one gas which is a source of acid rain.

_____ (1)

b) State one source of this gas.

_____ (1)

c) What environmental problems can occur if farmers use too much nitrate and phosphate fertiliser on their fields?

_____
_____
_____
_____ (2)

[4]

4 a) What is the main function of red blood cells?

_____ **(1)**

b) Identify two characteristics of a capillary and explain how each feature enables the capillary to perform its function.

_____

_____ **(4)**

c) State two harmful effects of smoking.

_____

_____ **(2)**

[7]

5 a) State two environmental conditions which are necessary for the production of oil.

_____

_____ **(2)**

b) The diagram shows how crude oil can be separated into useful products.

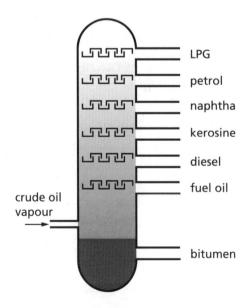

i) Name the process shown in the diagram.

_____ (1)

ii) The products from this separation are known as hydrocarbons. Which chemical elements do hydrocarbons contain?

_____ (2)

iii) Mark with an X the place where the hydrocarbons with the lowest boiling point will be collected. (1)

iv) Which gas is given off when hydrocarbons are burned as fuels?

_____ (2)

v) Why is it likely that renewable forms of energy will be increasingly used in the future?

_____ (2)

[10]

6 The alkali metals are: Lithium, Sodium, Potassium, Rubidium, Caesium, Francium.

a) Suggest which of these alkali metals:

i) has the highest melting point _____ (1)

ii) is the most reactive in water _____ (1)

iii) Explain your answer to ii) _____ (2)

b) Write a word equation to show the reaction of lithium and water.

_____ (2)

[6]

7 A student investigated the factors affecting the rate of reaction of marble chips with hydrochloric acid. In a series of experiments the student used whole and crushed chips, three different concentrations of acid and three different temperatures.

a) Suggest how the student could have measured the rate of this reaction.

_____ (1)

149

b) What is the likely effect of each of the following on the rate of reaction?

   i) Crushing the chips? _____ (1)

   ii) Warming the acid? _____ (1)

   iii) Using less concentrated acid? _____ (1)

c) How can the rate of gaseous reactions be increased without increasing the temperature or using the catalyst?

   _____ (1)

d) Part of the equation for the decomposition of hydrogen peroxide is shown below.

$$2H_2O_2 \xrightarrow{\text{catalyst x}} 2H_2O + \text{gas y}$$

   Name:

   i) catalyst x _____ (1)

   ii) gas y _____ (1)

   [7]

8 The diagram below shows a simplified diagram of a gas power station.

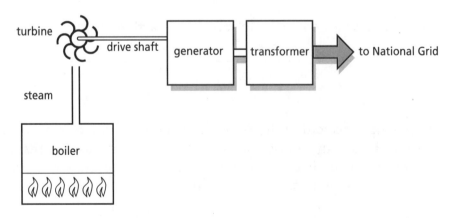

In which part of the power station:

a) is chemical energy converted to heat energy?

_____ (2)

b) is kinetic energy converted to electrical energy?

_____ (2)

c) Inside the generator, the drive shaft rotates a powerful magnet inside a coil of wire.

Why is it important that the magnet is rotating, not stationary?

_____
_____ (2)

d) What does the transformer do?

_____
_____ (2)

[8]

9 a) The diagram below shows a circuit used to run the two lights on a bike from one battery:

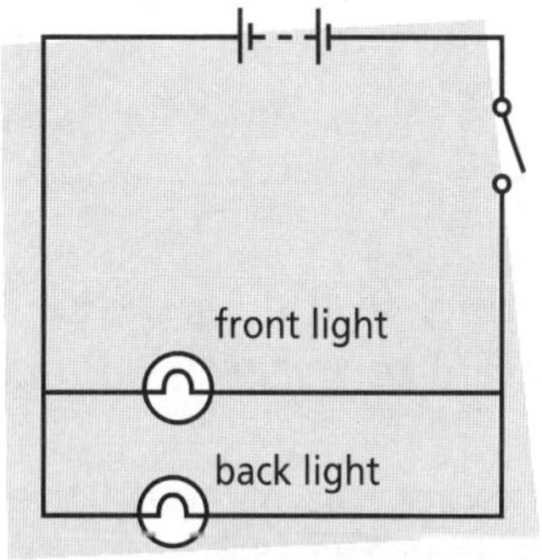

i) What is the name given for this way of connecting two bulbs together?

_____ (1)

ii) If the back light bulb blows, current can no longer flow through that bulb. Will the front light stay on or go out?

_____ (1)

b) An alternative way of connecting the bulbs is shown below.

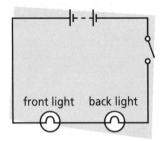

In this case, what happens to the front light bulb when the back bulb blows?

_____ (1)

[3]

10 The diagram shows the Earth and the Sun (not to scale).

a) There is a force acting on the Earth due to the Sun's gravity. Draw an arrow on the diagram to show the direction of this force. (1)

b) If the Sun did not exert this force on the Earth, what difference would this make to the Earth's movement?

_____

_____

_____ (2)

c) Consider the following four types of waves:

      light   infrared   ultraviolet   sound

Which one of these waves:

i) cannot travel to us from the Sun? _____ (1)

ii) makes us feel warm when the Sun is out? _____ (1)

iii) can cause skin cancer? _____ (1)

[6]

Total marks for paper = 58

Answers

1 a) i) Photosynthesis.
   ii) Carbon dioxide.
   b) Energy.

2 a) Kilojoules.
   b) Stomach.
   c) Undigested food lost as faeces / food used to provide energy / used up in respiration.

3 a) Carbon dioxide / sulphur dioxide / nitrogen dioxide.
   b) Carbon dioxide: respiration / combustion of fossil fuels.
   Sulphur dioxide: volcanoes / fossil fuel combustion.
   Nitrogen dioxide: fossil fuel combustion.
   c) Eutrophication / algal blooms.

4 a) Carry oxygen.
   b) Thin walls allow rapid diffusion of useful products and wastes / pores (holes) in walls allow fluid to leave blood.
   c) Smoking destroys cilia / causes build-up of mucus / increases blood pressure, increases risk of heart disease / attack.

5 a) High temperature / high pressures / long time periods.
   b) i) Fractional distillation.
   ii) Carbon and hydrogen.
   iii) Top of the column.
   iv) Carbon dioxide and water vapour.
   v) Fossil fuels are finite / will run out.

6 a) i) Lithium.
   ii) Caesium.
   iii) Reactivity related to distance of outer electron from nucleus / caesium has outer electron furthest from nucleus, therefore easiest to remove, therefore most reactive.
   b) Lithium + water → Lithium hydroxide + hydrogen.

7 a) Record time to collect 10cm³ of gas.
   b) i) Increases the rate.
      ii) Increases the rate.
      iii) Decreases the rate.
   c) Increase the pressure.
   d) i) Manganese IV oxide (manganese dioxide).
      ii) Oxygen.

8 a) Boiler.
   b) Generator.
   c) Rotating the magnet gives a changing magnetic field, which induces a voltage.
   d) Increases the voltage of the electricity leaving the power station.

9 a) i) In parallel.
      ii) It stays on.
   b) It goes out.

10 a)

   b) The Earth would not orbit the sun, but would go in a straight line into space.
   c) i) Sound.
      ii) Infrared.
      iii) Ultraviolet.